南京信息工程大学教材建设基金资助项目

U0150838

Laboratory Experiments for Inorganic and Analytical Chemistry

无机及分析化学实验

主 编 郭 彦

副主编 李 俊 徐 静 Srotoswini Kar 周永慧

特配电子资源

- 配套课件
- 视频学习
- 拓展阅读

南京大学出版社

图书在版编目(CIP)数据

无机及分析化学实验 ＝ Laboratory Experiments
for Inorganic and Analytical Chemistry / 郭彦主编
. — 南京：南京大学出版社，2021.4
ISBN 978 - 7 - 305 - 24357 - 8

Ⅰ. ①无… Ⅱ. ①郭… Ⅲ. ①无机化学－化学实验－
高等学校－教材②分析化学－化学实验－高等学校－教材
Ⅳ. ①O61－33②O65－33

中国版本图书馆 CIP 数据核字(2021)第 060501 号

出版发行　南京大学出版社
社　　址　南京市汉口路 22 号　　　　邮　编　210093
出 版 人　金鑫荣
书　　名　**无机及分析化学实验**
Laboratory Experiments for Inorganic and Analytical Chemistry
主　　编　郭　彦
责任编辑　刘　飞　　　　　　　编辑热线　025 - 83592146
照　　排　南京南琳图文制作有限公司
印　　刷　广东虎彩云印刷有限公司
开　　本　787×1092　1/16　印张 14.25　字数 325 千
版　　次　2021 年 4 月第 1 版　2021 年 4 月第 1 次印刷
ISBN 978 - 7 - 305 - 24357 - 8
定　　价　42.00 元

网址：http://www.njupco.com
官方微博：http://weibo.com/njupco
官方微信号：njupress
销售咨询热线：(025) 83594756

Preface

Under the condition of globalization, it is the foundation of higher education to cultivate talents with international vision, who can participate in international affairs and competitions. Experiment is an important part of chemistry: it is not only the basis and key for learning chemistry, but also it plays a special role in training students' practical ability and innovation ability. At present, imported English chemical experiment textbooks are expensive, and usually they adopt British-American system of units. At the same time, domestic English chemical experimental textbooks are very few, and there is no room for choice. These give us the motive force to publish this textbook.

This textbook is guided by the spirit of the National Conference on Undergraduate Education in Colleges and Universities in the new era and the concept of new discipline and specialty construction, based on the characteristics of domestic chemical experiment teaching, absorbing and learning from the experimental curriculum system and experimental contents of world-class universities. This textbook also keeps up with the pace of experimental teaching reform, and pays attention to the cultivation of innovation ability and interest in teaching objects.

Some contents of the teaching materials are directly taken from the self-made experimental handout of Reading Academy of Nanjing University of Information Science and Engineering, which has been used for many years. In the process of compiling this textbook, the accumulated teaching experience and self compiled handouts were further expanded, optimized and standardized, and officially published. Each experiment in this textbook is carefully selected, covering the basic experimental operation training to material synthesis, material identification, and then to qualitative and quantitative analysis. These experimental contents not only deepen and strengthen the theoretical knowledge of inorganic and analytical chemistry, but also consolidate students' experimental operation training.

Here, special thanks to those who paid painstaking effort for the publication of this textbook. Due to our limitation, omissions and improper points inevitably may occur. Please comment and correct it.

Content

1　INTRODUCTION

1.1　Laboratory Rules

Chemistry laboratories contain certain inherent dangers and hazards. As a chemistry student working in a laboratory, you must learn how to work safely with these hazards in order to prevent injury to yourself and others around you. You must make a constant effort to think about the potential hazards associated with what you are doing, and to think about how to work safely to prevent or minimize these hazards as much as possible. The following guidelines are here to help you. Please understand and follow these guidelines and act according to the principles behind them to help everybody as safe as possible. Ultimately, your own safety is your own responsibility. Please make sure you are familiar with the safety precautions, hazard warnings and procedures of the experiment you are performing on a given day before you start any work. If you are unsure of how to do something safely, please ask the teacher before proceeding. Experiments should not be performed without an instructor in attendance and must not be left unattended while in progress. No unauthorized experiments are allowed. No modification of the experiments is allowed. No work outside of regular hours is allowed, except under exceptional circumstances.

I. The general lab rules

(1) Make sure you are familiar with all the safety information given to you about each experiment before starting the experiment. This includes your manual, these safety guidelines, any posted information or any other information provided by your teacher.

(2) Always wear safety glasses (including during check-in and check-out), except when their removal has been specifically authorized by the teacher prior. Contact lenses are forbidden. You must also wear a face shield when requested by the teacher.

(3) You must wear a lab coat (and do it up) in all Chemistry labs.

(4) Footwear must completely cover the feet and heels (no sandals, baby dolls, ballet flats, mules, open-toed footwear, etc.).

(5) You must wear long pants (no shorts, capris, skirts or dresses).

(6) If you arrive at your Chemistry lab and do not have the required clothing,

you will be directed to rent or purchase missing items (glasses, lab coats, disposable footcoverings and long pants) from Laboratory Center before you will be allowed to participate in the lab.

(7) Loose hair must be tied back so as to be out of the way. Dangling jewellery must be removed.

(8) Do not eat or drink in the lab.

(9) Visitors are not allowed to be in the lab.

(10) Please keep your work area and the common work areas tidy. Also, please make sure the aisles, safety showers, eyewash stations and doorways are unobstructed.

(11) Please leave all the glassware, equipment, tools, etc. as clean as or cleaner than you found them.

(12) Please clean up spills immediately. If the spill is large or is of a hazardous material, inform the teacher immediately. Use spill mix to absorb solvent or caustic liquids.

(13) Please dispose of waste properly and in a timely manner and according to the instructions provided in your lab manual. If you are not sure, please ask your teacher for the proper method of disposal.

(14) Wash your hands before you leave the lab.

(15) Do not remove chemicals or equipment from the lab except when required to do so for analysis.

(16) Please notify your teacher of any serious medical conditions.

(17) Do not wear earbuds or earphones while in the lab.

II. Handling reagents and standard procedures

The liquids, solids and solutions used in a laboratory are called reagents. You must become well acquainted with these reagents, their containers, and their proper use. The reagents are kept on a separate bench or hood away from your work area. Some reagents must be kept in the fume hood because they generate flammable or toxic fumes. The reagents are grouped according to experiment, when you need a reagent please follow these rules:

(1) Be sure to use the correct reagent. Before using the reagent, carefully check the chemical name, formula, concentration, and double check to be sure you have the right one. Note the hazard code and warnings and take necessary precautions.

(2) Do not take reagent containers to your work area, and take only what you need.

(3) Do not contaminate the reagents. Always use a clean spatula for solids and clean glassware for liquids. Never put a pipet or pipettor into a liquid reagent, instead pour what is needed into a clean, dry container and take it to your work area to pipet

from there.

(4) Put the lids back on the reagent containers snugly and put themback in the correct locations. Clean up any reagents you spill with a wet sponge, rinse out the sponge at the sink, and then wash your hands.

(5) Never return unused reagents, liquid or solid, to the reagent bottles. Discard or share any excess. Label any container you use to store a reagent with the chemical name and hazard or hazard code. The concentration and chemical formula along with your name, section, and date would also be good information to add to the label.

(6) Use great care with corrosive chemicals (strongly acidic or basic solutions). Always wear safety goggles! Rinse your hands with tap water after using corrosive chemicals, especially if you feel a burning or slimy sensation on your skin. Wear the gloves provided in the laboratory if called for. Most strong acids and bases will be disposed of in the "Corrosive Liquids bucket", as noted in experimental procedures unless the used chemical has other hazardous properties.

(7) Dispose of nonhazardous chemicals in the large sinks available in the lab. Be sure to follow the instructions in the experiments with regard to the disposal of chemicals.

(8) Pure water (PW) is made using activated carbon filtration, reverse osmosis (RO), and ion exchange or distillation followed by UV treatment to remove any salts or organic compounds and kill any microbes that could contaminate your solutions. All pure water taps will be labeled with PW. When washing glassware, often all that is needed is to rinse well with tap water 3 or 4 times followed by one rinse with PW inside and out. If the glassware is really dirty use detergent, then rinse tap water. Then, rinse all glassware with PW from a wash bottle or carboy filled with PW before use or storage. Fill your plastic wash bottle with PW for this purpose. You do not need to dry the inside of the glassware. Never store dirty glassware.

(9) At the end of every lab period you must clean your workstation bench space and any area you used by wiping it down with a clean, damp sponge. Rinse out and wring out the sponge when you are done. Your workstation drawer must be neat and complete with clean glassware and equipment for the next student. If you break glassware during lab, be sure to obtain a replacement from your teacher before you leave. Do not store your goggles, solutions, or unknowns in your workstation. Instead place them in your student storage bin.

1.2 Laboratory Safety

Faulty technique is one of the chief causes of accidents, because it involves the human element, and is one of the most difficult to cope with. The purpose of this discussion is to help the student understand proper laboratory safety, to increase his awareness of the possible risks or hazards involved with laboratory work and to realize the laboratory is generally a safe place to work if safety guidelines are properly followed.

I. Standard operating procedures

A. General personal safety

(1) Eating, drinking, smoking, applying cosmetics or lip balm, and handling of contact lenses are prohibited in areas where specimens are handled.

(2) Food and drink should not be stored in refrigerators, freezers, cabinets, or on shelves, countertops, or bench tops where blood or other potentially infectious materials are stored or in other areas of possible contamination.

(3) Long hair, ties, scarves and earrings should be secured.

(4) Keep pens and pencils out of your mouth.

(5) Appropriate Personal Protective Equipment (PPE) will be used when indicated:

Lab coats or disposable aprons should be worn in the lab to protect you and your clothing from contamination. Lab coats should not be worn outside the laboratory.

Lab footwear should consist of normal closed shoes to protect all areas of the foot from possible puncture from sharp objects and/or broken glass and from contamination from corrosive reagents and/or infectious materials.

Gloves should be worn for handling blood and body fluid specimens, touching the mucous membranes or non-intact skin of patients, touching items or surfaces soiled with blood or body fluid, and for performing venipunctures and other vascular access procedures. Cuts and abrasions should be kept bandaged in addition to wearing gloves when handling biohazardous materials.

Protective eyewear and/or masks may need to be worn when contact with hazardous aerosols, caustic chemicals and/or reagents is anticipated.

(6) Never mouth pipette. Use your hand or mechanical pipetting devices for pipetting all liquids.

(7) Frequent hand washing is an important safety precaution, which should be practiced after contact with patients and laboratory specimens. Proper hand washing techniques include soap, running water and 10—15 seconds of friction or scrubbing action. Hands should be dried and turn the faucets off with the paper towe.

Hands are washed: ① After completion of work and before leaving the laboratory. ② After removing gloves. ③ Before eating, drinking, smoking, applying cosmetics, changing contact lenses or using lavatory facilities. ④ Before all other activities which entail hand contact with mucous membranes or breaks in the skin. ⑤ Immediately after accidental skin contact with blood or other potentially infectious materials. ⑥ Between patient contact and before invasive procedures.

(8) Laboratory work surfaces must be disinfected daily and after a spill of blood or body fluid with a 1 : 10 dilution of clorox in water.

B. Eye safety

(1) Know where the nearest eye wash station is located and how to operate it.

(2) Eye goggles should be worn: ① When working with certain caustic reagents and/or solvents, or concentrated acids and bases. ② When performing procedures that are likely to generate droplets/aerosols of blood or other body fluid. ③ When working with reagents under pressure. ④ When working in close proximity to ultra-violet radiation (light).

(3) Wearing contact lenses in the laboratory is discouraged and requires extra precaution if worn. Gases and vapors can be concentrated under the lenses and cause permanent eye damage. Furthermore, in the event of a chemical splash into an eye, it is often nearly impossible to remove the contact lens to irrigate the eye because of involuntary spasm of the eyelid. Persons who must wear contact lenses should inform their supervisor to determine which procedures would require wearing no-vent goggles.

C. Safe handling of biologically hazardous material

(1) You should handle all patient samples as potentially biohazardous material. This means universal precautions should be followed at all times.

(2) When working in the laboratory: ① Wear protective clothing (lab coat, gloves. If you have a cut/abrasion, also wear a band-aid. ② Avoid spillage and aerosol formation. ③ Hands should be washed immediately and thoroughly if contaminated with blood or other body fluids. ④ Gloves should be removed before handling a telephone, computer keyboard, etc. , and must not be worn outside the immediate work area. ⑤ Hands should always be washed immediately after gloves are removed. ⑥ You should wash your hands after completing laboratory activities and before leaving the area. All protective clothing should be removed prior to leaving the lab. ⑦ All biohazardous material should be discarded in a biohazard bag to be autoclaved. ⑧ All counter and table tops should be disinfected with a proper disinfecting solution at the beginning and the end of the day, and when you should spill a patient sample.

(3) Proper handling of sharps: ① Contaminated needles and other sharps are never broken, bent, recapped or re-sheathed by hand. ② Used needles are not removed from disposable syringes. ③ Needles and sharps are disposed of in impervious containers located near the point of use.

II. Chemical and gas safety

To provide a safe working environment, all personnel should be aware of potentially hazardous materials and the proper way of handling this material. Avoid unnecessary exposure to chemicals. Material Safety Data Sheets(MSDS) concerning the handling of hazardous materials should be available to all laboratory personnel, so that they may achieve and maintain safe working conditions.

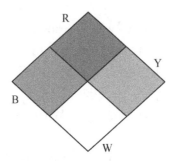

Flammable (Red); Instability (Yellow); Health (Blue) and Special Notice (White)

Figure 1 - 1   NFPA chemical hazard sign

The positions on the NFPA diamond are defined as follows:

Health Hazard (Blue): Degree of hazard for short-term protection.

0. Ordinary combustible hazards in a fire

1. Slightly hazardous

2. Hazardous

3. Extreme danger

4. Deadly

Flammability (Red): Susceptibility to burning.

0. Will not burn

1. Will ignite if preheated

2. Will ignite if moderately heated

3. Will ignite at most ambient conditions

4. Burns readily at ambient conditions

Reactivity, Instability (Yellow): Energy released if burned, decomposed, or mixed.

0. Stable and not reactive with water

1. Unstable if heated

2. Violent chemical change

3. Shock and heat may detonate

4. May detonate

Special Hazard (White position ondiamond):

OX: Oxidizer

W: Use no water, reacts!

Apart from requiring that MSDS be available to workers, one of the other important aspects is the requirement for clear labels and hazard symbols on hazardous products. The following eight hazard symbols should be used as guides for the handling of chemical reagents:

Symbol	Class Description	Symbol means that the material:
	Compressed Gas (Class A)	• Poses an explosion danger because the gas is being held in a cylinder under pressure • may cause its container to explode if heated • may cause its container to explode if dropped
	Combustible and Flammable Material (Class B)	• is one that will burn and is consequently a fire hazard (i. e. , is combustible) • may ignite spontaneously in air or release a flammable gas in contact with water
	Oxidizing Material (Class C)	• may react violently or cause an explosion when it comes into contact with combustible materials • may burn skin and eyes upon contact
	Poisonous Material: Immediate Toxic Effects (Class D1)	• is a potentially fatal poisoning substance • may be immediately fatal or cause permanent damage if it is inhaled or swallowed or enters the body through skin contact
	Poisonous Material: Other Toxic Effects (Class D2)	• is a poisonous substance that is not immediately hazardous to health • may cause death or permanent damage as a result of repeated exposure over time (e. g. , cancer, birth defects or sterility) • may be an irritant
	Biohazardous Infectious Material (Class D3)	• may cause a serious disease resulting in illness or death • may produce a toxin that is harmful to humans
	Corrosive Material (Class E)	• causes severe eye and skin irritation upon contact • causes severe tissue damage with prolonged contact
	Dangerously Reactive Material (Class E)	• is very unstable • may react with water to release a toxic or flammable gas • may explode as a result of shock, friction, or increase in temperature • may explode if heated in a closed container

A. Compressed gases

(1) The storage of all compressed gases shall be in containers designed, constructed, tested and maintained in accordance with the Gas Cylinder Use and Management Regulations.

(2) In the laboratory, gas containers are to be limited to the number of containers in use at any time. Low pressure (LP) gases shall also be limited to the smallest size container.

(3) Containers shall be securely strapped, chained or secured in a cylinder stand so they cannot fall.

(4) Oxidizing gases should be separated from flammable gases.

B. Flammable compounds

(1) All flammable reagents should be kept in the flammable storage facilities (closet or refrigerator) at all times when not in use.

(2) Any solutions compounded from these reagents should be labeled as flammable.

(3) Flammable substances should be handled in areas free of ignition sources.

(4) Flammable substances should never be heated using an open flame.

(5) Ventilation is one of the most effective ways to prevent accumulation of explosive levels of flammable vapors. An exhaust hood should be used whenever appreciable quantities of flammables are handled.

(6) Flammable compounds should be placed in proper receptacle for disposal.

C. Ether precautions (oxidizing and flammable compounds)

(1) These compounds tend to react with oxygen to form explosive peroxides. When ether containers are opened they are to be dated and all material remaining after six (6) months must be disposed of immediately.

(2) Disposal of ether compounds is through the Hazardous Materials Office.

(3) Ether compounds will be stored in an explosion-proof refrigerator. (boiling point of ether is approximately room temperature)

D. Toxic and corrosive materials (acids and alkali)

(1) To avoid dangerous splatter, always add acid to water!

(2) Toxic materials should be labeled with special tape when used in compounded reagents and stored in separate containers. These materials should be handled carefully and kept in the hood during preparation.

(3) Acids and alkali should be carried by means of special protective carriers when transported.

(4) Acid and alkali spills should be covered and neutralized by using the material from the 'spill bucket'. All material, spill and compound, should be swept up and placed in a plastic bucket for proper disposal.

（5）In case of spillage, wash all exposed human tissue （including eyes） generously with water and notify your supervisor for proper reporting of the incident.

E. Carcinogens

（1）All laboratory chemicals identified as carcinogens must be labeled carcinogen.

（2）When working with these substances, protective clothing and gloves should be worn.

III. Radiation safety

A. No eating, drinking, smoking permitted!

B. Radioactive material should be labeled as radioactive and stored in a proper container so as to prevent spillage or leakage.

C. These materials must be handled carefully. Remember: the amount of radiation exposure decreases with distance.

D. Radioactive spills should be absorbed with absorbent toweling. The area should be cleaned with soap and water and then decontaminated with a product such as 'count-off'. The area of the spill is then monitored for any residual radioactivity. If the area is not decontaminated, the above regimen is repeated and re-monitored.

E. In the case of a radioactive spill in a high traffic area, the area will be 'roped off' until proper decontamination has been achieved.

F. In the case of a major radioactive spill, all personnel in the area must be notified. The appropriate safety officer must be notified and all attempts to keep contamination at a minimum must be used.

IV. Fire safety

A. Know where all fire exits, fire extinguishers and fire alarms are located!

B. Know how to properly operate appropriate fire alarms and fire safety equipment!

C. Know the proper procedure for notifying colleagues and proper personnel of a fire.

If it is safe for you to attempt to extinguish the fire, remember R-A-C-E-R:

（1）Rescue those in danger.

（2）Alarm: ① Activate the fire pull station. ② Notify supervisors of the location, your name and the type of fire, if known.

（3）Contain the fire by closing all doors and windows.

（4）Extinguish the fire, if possible. Do not re-enter a room that has already been closed.

（5）Relocate: Evacuate the building, if necessary.

V. Electrical safety

A. The use of extension cords is prohibited.

B. All equipment must be properly grounded.

C. Never operate electrical equipment with fluid spillage in the immediate area or with wet hands.

D. Never use plugs with exposed or frayed wires.

E. If there are sparks or smoke or any unusual evens occur, shut down the instrument and notify the manager or safety officer. Electrical equipment that is not working properly should not be used.

F. If a person is shocked by electricity, shut off the current or break contact with the live wire immediately. Do not touch the victim while he is in contact with the source of current unless you are completely insulated against shock. If the victim is unconscious, call 110 to report the incident and request assistance.

VI. Severe weather safety

A. When the tornado-warning message is heard on the public address system, all personnel should move to a safe area. Safe areas are considered to be: (1) Below ground level if possible. (2) Inside, interior halls in an east/west corridor, away from windows. (3) Inside, interior windowless rooms.

B. Stair towers should be used for evacuation.

C. Elevators should be used only in emergency.

D. No one will leave the building until the 'all clear' is announced.

VII. General procedures and equipment

A. Cracked or chipped glassware should not be used.

B. Centrifuges should not be used without the covers completely closed.

C. When removing tops from evacuated test tubes, care must be taken to prevent aerosol formation.

VIII. In case of accidents

A. Accidental Needle Stick: (1) Bleed wound. (2) Wash wound thoroughly with soap. (3) Notify the supervisor of the incident and report to Student Health with an incident report form. (4) May need to get blood tested for hepatitis.

B. If you should wound yourself in the laboratory: Any type of accident should be brought to the attention of the Teaching Supervisor of the area.

1.3 Laboratory Common Equipments

Beaker

（Container, reaction vessel, Can roughly measure the volume）

Conical beaker

（Container, reaction vessel, widely used in titration reaction）

Round bottom flask

（Reaction vessel, usually used with other equipment）

3 neck Round bottom flask

（Reaction vessel, usually used with other glass ware）

Volumetric or graduated flask

（one accurate volume only）

Buchner flask

（for vacuum filtration）

Pear shaped flask

（for rotary evaporation）

round

Separating flask pear

（for separating or adding liquid）

straight

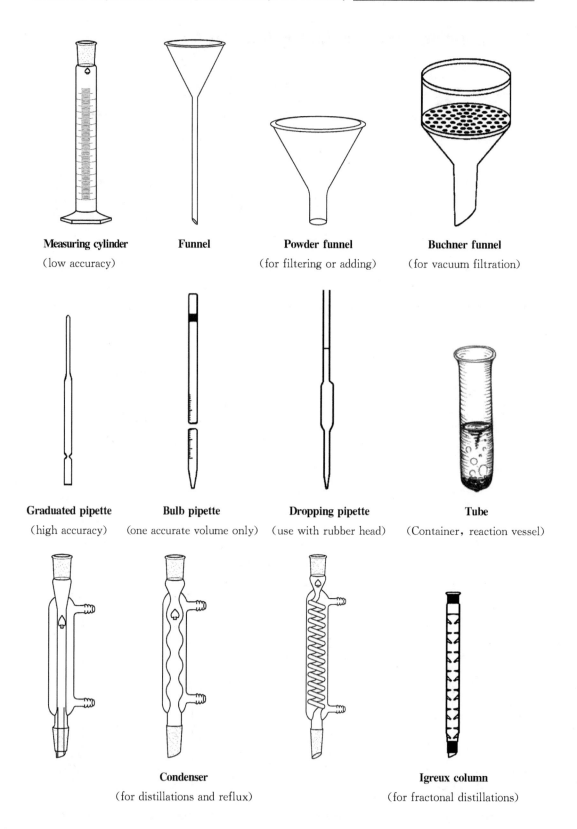

Measuring cylinder

(low accuracy)

Funnel

Powder funnel

(for filtering or adding)

Buchner funnel

(for vacuum filtration)

Graduated pipette

(high accuracy)

Bulb pipette

(one accurate volume only)

Dropping pipette

(use with rubber head)

Tube

(Container, reaction vessel)

Condenser

(for distillations and reflux)

Igreux column

(for fractonal distillations)

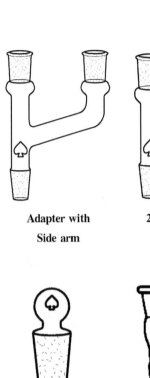

Adapter with
Side arm

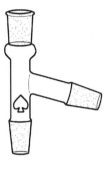

2 Neck Adapter

Dropping or
addition funnel

Claisen flask with
Vigreux column

Stopper

Expansion adapter

Reduction adapter

Boss

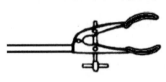

Clamp

Ring clamp

Drying tube

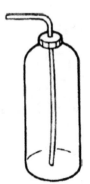

Washing bottle

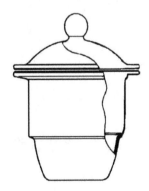

Desiccator
(keep samples to prevent
water absorption)

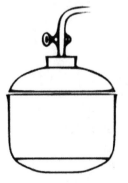

Desiccator with Stopcock
(use under vacuum)

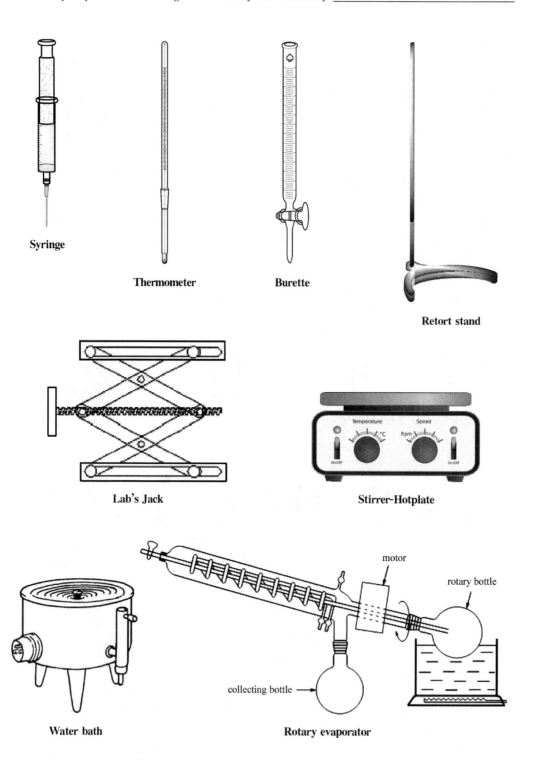

Syringe

Thermometer

Burette

Retort stand

Lab's Jack

Stirrer-Hotplate

motor

rotary bottle

collecting bottle

Water bath

Rotary evaporator

1.4 Pre-lab Work

Chemistry lab is a required component of most chemistry courses. Learning about lab procedures and performing experiments helps you to learn techniques and reinforces textbook concepts. Make the most of your time in the lab by coming to the lab prepared.

These pre-lab tips will help you before starting an experiment.

(1) Complete any pre-lab assignments or homework. The information and calculations are intended to make the lab exercise quicker and easier.

(2) Know the location of the lab safety equipment and understand how to use it. In particular, know the location of the emergency exit, fire extinguisher, eyewash station, and safety shower.

(3) Read through the experiment before going to the lab. Make sure you understand the steps of the experiment. Jot down any questions you have so that you can ask them before starting the lab.

(4) Start filling out your lab notebook with information about the experiment. It's a good idea to draw out your data table in advance so all you need to do in the lab is fill it in with numbers.

(5) Review the Material Safety Data Sheets (MSDSs) of the chemicals you will be using during the lab.

(6) Make certain you have all of the glassware, materials, and chemicals needed to complete the lab before starting any part of the procedure.

(7) Understand the disposal procedures for the chemicals and other items used in your experiment. If you are unclear about what to do with your experiment after it has been completed, ask your instructor about it. Don't throw items in the trash or dump liquids down the drain or in waste disposal containers until you are certain it is acceptable to do so.

(8) Be prepared to take data in the lab. Bring your notebook, a pen, and a calculator.

(9) Have personal safety gear, such as a lab coat and goggles, clean and ready to use before the lab.

1.5 Lab Reports

Lab reports are an essential part of all laboratory courses and usually a significant part of your grade. If your instructor gives you an outline for how to write a lab report, use that. Some instructors require the lab report be included in a lab notebook, while others will request a separate report. Here's a format for a lab report you can use if you aren't sure what to write or need an explanation of what to include in the different parts of the report.

A lab report is how you explain what you did in your experiment, what you learned, and what the results meant.

Lab Report Essentials:

(1) Title page

Not all lab reports have title pages, but if your instructor wants one, it would be a single page that states:

- The title of the experiment.
- Your name and the names of any lab partners.
- Your instructor's name.
- The date the lab was performed or the date the report was submitted.

(2) Title

The title says what you did. It should be brief (aim for ten words or less) and describe the main point of the experiment or investigation. If you can, begin your title using a keyword rather than an article like 'The' or 'A'.

(3) Introduction/Purpose

Usually, the introduction is one paragraph that explains the objectives or purpose of the lab. In one sentence, state the hypothesis. Sometimes an introduction may contain background information, briefly summarize how the experiment was performed, state the findings of the experiment, and list the conclusions of the investigation. Even if you don't write a whole introduction, you need to state the purpose of the experiment, or why you did it. This would be where you state your hypothesis.

(4) Materials

List everything needed to complete your experiment.

(5) Methods

Describe the steps you completed during your investigation. This is your procedure. Be sufficiently detailed that anyone could read this section and duplicate your experiment. Write it as if you were giving directions for someone else to do the lab. It may be helpful to provide a figure to diagram your experimental setup.

(6) Data

Numerical data obtained from your procedure usually is presented as a table. Data encompasses what you recorded when you conducted the experiment. It's just the facts, not any interpretation of what they mean.

(7) Results

Describe in words what the data means. Sometimes the Results section is combined with the Discussion (Results & Discussion).

(8) Discussion or analysis

The Data section contains numbers. The Analysis section contains any calculations you make based on those numbers. This is where you interpret the data and determine whether or not a hypothesis was accepted. This is also where you would discuss any mistakes you might have made while conducting the investigation. You may wish to describe ways the study might have been improved.

(9) Conclusions

Most of the time the conclusion is a single paragraph that sums up what happened in the experiment, whether your hypothesis was accepted or rejected, and what this means.

(10) Figures and graphs

Graphs and figures must both be labeled with a descriptive title. Label the axes on a graph, being sure to include units of measurement. The independent variable is on the X-axis. The dependent variable (the one you are measuring) is on the Y-axis. Be sure to refer to figures and graphs in the text of your report. The first figure is Figure 1, the second figure is Figure 2, etc.

(11) References

If your research was based on someone else's work or if you cited facts that require documentation, then you should list these references.

1.6 Data Analysis

I. Recording experimental measurements

A report sheet is used as well as a laboratory notebook to provide more structure in data collection. You should be able to look at your lab notebook a year from now and be able to repeat the experiment or calculations. Calculations should be shown in the lab notebook. Dimensional analysis (unit cancellation) must be used to do all the calculations in this course. If a spreadsheet or graph is used to do the calculations, staple a copy to the report sheet and lab notebook pages.

Before you start an experiment, key aspects of the laboratory procedure should be outlined in your lab notebook. Any procedures not in the lab manual including changes to procedures listed in the manual must be noted. Key data must also be recorded in the laboratory notebook, in case the report sheet is lost. Certain rules need to be followed when keeping a lab notebook:

(1) Record all data and observations directly in the lab notebook. This is by far the most important rule in recording data. Do not transcribe data from other pieces of paper, i. e. , do not record data on scraps of paper and then recopy the data into the

lab notebook. Write down exactly what you are doing and your observations as you are doing the experiment. Errors in your procedure can be caught this way. Points can be taken off for writing raw data in places other than the lab notebook/report sheet.

(2) Clearly identify all data, graphs, axes, and use correct units. Use unit cancellation.

(3) A ball point pen must be used for all entries in a lab notebook. A pen must be used for all measured data (mainly mass and volume data) and observations. Do not white out, erase any entry; simply cross out mistakes with a single line (the mistake should still be readable) and give a short note to explain the nature of the mistake, e. g. , "misread." Sometimes you will find later that the entry was not a mistake after all and will want to retrieve the data. So never obliterate or destroy data no matter how bad it looks!

(4) Before an experiment is started, the entire experimental procedure must be read. As you read it, note the objectives and key points of the experimental procedure in your lab notebook. This will prepare you for the pre-lab quiz and experiment before you come to the lab.

(5) Another important facet of scientific experiments involves the propagation of accuracy (or inaccuracy) of measurements through the calculations to the results. Use the correct number of significant figures, as outlined below, during the collection of data and calculations.

II. Recording experimental data using correct significant figures

It is important to take data and report answers such that both the one doing the experiment and the reader of the reported results know how precise the results are. The simplest way of expressing this precision is by using the concept of significant figures. A significant figure is any digit that contributes to the accuracy of an experimentally measured number or to a number calculated from experimentally measured numbers. Please refer to the chemistry lecture textbook for a discussion pertaining to the use of significant figures.

Usually, mass, volume, time, and temperature are experimentally measured and used to calculate density, concentration, percent by mass, and other values of interest. For example, mass in grams (g) is always measured using a top loading electronic balance with a precision of ± 0.001 g. Most mass measurements should be recorded to this precision even though the last digit may vary somewhat. For example, if the mass of an object on a balance reads 25. 001, 25. 000, 24. 999 and moves between these values, 25. 000 should be recorded. Recording 25, 25. 0, or 25. 00 would be wrong since these would not communicate the true precision of the number. If values on the balance change randomly from 25. 000, 25. 001, to 25. 002

then 25.001 g should be recorded. For very precise mass measurements an analytical balance is used to ±0.000 1 g.

Time in seconds (s) is measured using a timer, stopwatch, or perhaps a clock so the precision of the measurement might vary from ±1 to ±0.01 seconds. Always record the number to the maximum precision. Temperature will be measured using an alcohol thermometer that can be read to a precision of ±0.2 ℃ so estimate to the tenth of a degree (i.e. 21.3 ℃).

Measuring volume in mL is a tradeoff between speed and the precision of the measurement and requires skill in choosing the right glassware for the task. When an approximate volume is needed, a beaker, Erlenmeyer flask, or graduated cylinder can be used, but when an accurate volume is needed, a pipet, pipettor, buret, or volumetric flask will be specified for use. Recognizing when to make an accurate measurement and when to be satisfied with an approximate measurement can save much time.

Frequently, the written directions will give clues to the needed precision by using the words "approximately" or "about" when the precision is not important and "exactly" or "precisely" when the precision is important. Another clue would be the number of significant figures used to write a number. It is also important to note that glassware used for accurate measurements is calibrated at a specific temperature, which is noted on the glassware. The precision of various types of glassware is shown in the following table 1-1.

Table 1-1　Precision of Glassware for Volume Measurement

Equipment	precision	Purpose of glassware/equipment
250 mL beaker	±10 mL	Solution preparation, storage, reactions
125 mL conical beaker	±6 mL	Solution preparation, storage, reactions
250 mL graduated cylinder	±1 mL	Volume transfer-moderate precision
25 mL graduated cylinder	±0.2 mL	Volume transfer-moderate precision
5 mL pump dispenser	±0.1 mL	Volume transfer-moderate precision
100 mL volumetric flask	±0.08 mL	Precise final volume for dilutions
10 mL measuring pipet	±0.05 mL	Volume transfer-good precision
5 mL pipettor	±0.025 mL	Volume transfer-very precise
25 mL buret	±0.02 mL	Precise volume delivery for titration
5, 10 mL volumetric pipet	±0.01 mL	Volumetransfer-very precise

When a measurement is made, the question arises: "How many digits or figures should be recorded?" The answer is straightforward: For a measured number records all digits, which are known with certainty, and the last digit, which is estimated.

Many of the measurements in this course involve estimation to the nearest one-fifth or one-tenth of a scale marking. For example, a 25 mL graduated cylinder, which has scale markings every 0. 5 mL, should be read to the nearest 0. 1 mL, estimation to the nearest one-fifth of a division. The graduated cylinder does not need to be used to this accuracy at all times; for example, if the instruction says "add about 25 mL of water" being within 1—2 mL of 25 would be ok.

NOTE: Whenever estimation between markings is being done and the reading is "on the mark," the last digit should be included to convey the idea of accuracy to the reader. For example, with a burette, which has markings every 0. 1 mL, a reading on the mark of 11. 3 mL would be recorded as 11. 30 mL; otherwise, the reader will not know that the burette was really read to the nearest 0. 01 mL. (You must estimate the last digit by looking carefully between the markings).

Sometimes approximate small amounts of liquid are needed. In this case instructions may indicate to measure out drops from a dropper bottle or eye dropper. One drop of water or a dilute solution on average is about 0. 05 mL. This can also be a safer method because it does not involve pouring the liquid from one container to another.

Generally speaking, all the glassware in the table on the previous page is for transferring known volumes of liquid from one container to another except for the beaker and flasks. Beakers along with conical beakers are generally used for conducting chemical reactions or other lab manipulations. The volumetric flask is used for preparing precise solutions or dilutions.

III. Interpretation of data

Significant figures are excellent to express the precision of raw data but not as good to express the precision of calculated values. As a general rule in this laboratory course you should always use at least four significant figures for calculated values to avoid rounding errors. In order to interpret the quality of your results, certain terms are useful. You will need to understand the following definitions.

(1) Accuracy: The term "accuracy" describes the nearness of a measurement to its accepted or true value.

(2) Precision: The term "precision" describes the "reproducibility" of results. It can be defined as the agreement between the numerical values of two or more measurements (trials) that have been made in an identical fashion. Good precision does not necessarily mean that a result is accurate.

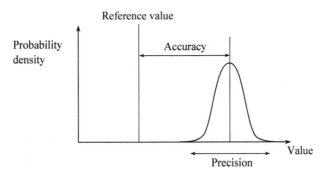

Figure 1 - 2 Probability curve about accuracy and precision

(3) Range: The "range" is one of several ways of describing the precision of a series of measurements. The range is simply the difference between the lowest and the highest of the values reported. As the range becomes smaller, the precision becomes better.

Example: Find the range of 10. 06, 10. 38, 10. 08, and 10. 12.

$$Range = 10.38 - 10.06 = 0.32$$

(4) Mean: The "mean" or "average" is the numerical value obtained by dividing the sum of a set of repeated measurements by the number of individual results in the set.

Example: Find the mean of 10. 06, 10. 38, 10. 08, 10. 12,

$$\frac{10.06 + 10.38 + 10.08 + 10.12}{4} = 10.16$$

(Note that the value 10. 38, which is far greater than the other values, has a large influence on the mean, which is larger than three out of the 4 individual values.)

(5) Median: The "median" of a set is that value about which all others are equally distributed, half being numerically greater and half being numerically smaller. If the data set has an odd number of measurements, selection of the median may be made directly.

Example: the median of 7. 9, 8. 6, 7. 7, 8. 0 and 7. 8 is 7. 9, the "middle" of the five.

For an even number of data, the average of the central pair is taken as the median.

Example: the median of 10. 06, 10. 38, 10. 08, and 10. 12 is 10. 10 which is the average of the middle pair of 10. 08 and 10. 12.

Notice in the example that the median is not influenced much by the value 10. 38, which differs greatly from the other three values as in the example for the mean above. For this reason, the median is usually better to use in reporting results than the mean for small data sets.

(6) Error: The absolute error of an experimental value is the difference between it and the true value. For example if the experimental value is 30. 9 and the known true value is 26. 5, the error would be 30. 9—26. 5 or 4. 4.

Systematic errors occur when there is an error in the experimental procedure. Measuring the volume of water from the top of the meniscus rather than the bottom, or overshooting the volume of a liquid delivered in a titration will lead to readings which are too high. Heat losses in an exothermic reaction will lead to smaller observed temperatures changes.

Random errors are caused by the readability of the measuring instrument, or the effects of changes in the surroundings, such as temperature variations and air currents, or insufficient data, or the observer misinterpreting the reading. Random errors make a measurement less precise, but not in any particular direction.

Random uncertainties can be reduced by repeating readings; systematic errors cannot be reduced by repeating measurements.

Relative percent error would be the error divided by the true value times 100%: (4. 4/26. 5)×100%=16. 6% or 17%.

(7) Standard deviation (SD, also represented by the Greek letter sigma σ or the Latin letter s) is a measure that is used to quantify the amount of variation or dispersion of a set of data values. In the case where x takes random values from a finite data set x_1, x_2, $\cdots$, x_n, with each value having the same probability, the standard deviation is

$$\sigma=\sqrt{\frac{1}{n}(x_1-\mu)^2+(x_2-\mu)^2+\cdots+(x_n-\mu)^2}$$

Where, μ is mean of the data set x_1, x_2, $\cdots$, x_n.

Relative standard deviation (RSD), is a standardized measure of dispersion of a probability distribution or frequency distribution. It is often expressed as a percentage, and is defined as the ratio of the standard deviation σ to the mean μ (or its absolute value, $|\mu|$)

IV. Graphing and Analyzing Data

Graphical techniques are an effective means of communicating the effect of an independent variable on a dependent variable, and can lead to determination of physical quantities. The independent variable is the cause and is plotted on the horizontal axis. The dependent variable is the effect and is plotted on the vertical axis.

Sketched graphs have labelled but unscaled axes, and are used to show qualitative trends, such as variables that are proportional or inversely proportional. Drawn graphs have labelled and scaled axes, and are used in quantitative measurements.

When drawing graphs:

(1) Give the graph a title and label the axis with both quantities and units.

(2) Use the available space as effectively as possible and use sensible scales-there should be no uneven jumps.

(3) Plot all the points correctly.

① Identify any points which do not agree with the general trend.

② Think carefully about the inclusion of the origin. The point (0, 0) can be the most accurate data point or it can be irrelevant.

A best-fit straight line does not have to go through all the points but should show the overall trend.

The equation for a straight line is:

$$y=sx+c$$

Where: x is the independent variable, y is the dependent variable, $s=\Delta y/\Delta x$, is the gradient and has units, c is the intercept on the vertical axis.

A systematic error produces a displaced line. Random uncertainties lead to points on both sides of the perfect line.

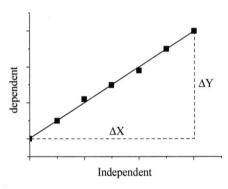

Figure 1 - 3　A best-fit of data points

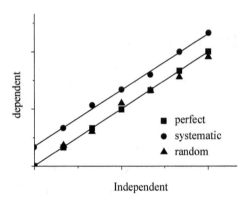

Figure 1 - 4　Systematic error and random error

2 EXPERIMENTS

2.1 Calibration of Volumetric Glassware

AIMS

1. Validation on the basic principles and fundament of chemistry, and then deepening and expanding the theoretical learning.

2. To master the basic skills of experimental operation.

3. To learn how to analyze and solve problems.

INTRODUCTION

This experiment is designed to introduce you to some of the apparatus and proper analytical laboratory techniques you will be using during the remainder of this course, including the methods for evaluating analytical data treatment. As a chemist, it is imperative to gather the best possible data from your equipment. Therefore, it is important to use all equipment correctly, patiently and precisely. In addition, a good chemist will calibrate an "instrument" before they use it to assure that the data gathered is as accurate as possible.

I. General guidelines for the analytical laboratory

(1) Never assume any glassware is clean unless you washed it yourself!!!

(2) Keep your bench area clean!

(3) Clean all the glassware with soap and tap water unless otherwise directed.

(4) Use distilled water to rinse your glassware, preferably a double or triple rinse.

(5) Glassware that will not come clean in soap and water should be cleaned by soaking in a solution of 1 M Nitric Acid or 1 M KOH in alcohol, followed by a soap and water wash and distilled water rinse. This is typically true of Burettes.

(6) Never heat any piece of volumetric glassware. Heating will cause changes in

the glassware that will cause it not to be volumetric anymore.

(7) All items in this laboratory will be massed using the analytical balance. These balances read to the nearest 0. 1 mg (0. 000 1 g). Thus, all masses recorded must be to the nearest 0. 1 mg. (Do not record anything from the balance without all 4 places)

(8) You will be assigned a balance to use. You will use the same analytical balance for all measurements throughout the semester. This improves the precision of your measurements. (Plus I know who to penalize when it is trashed!)

(9) You must clean the analytical balance when you are finished with it! Failure to do so will result in a 5 point deduction from the laboratory reports of all the students assigned to that balance.

(10) Burettes must be washed at the end of every laboratory period (soap, water, rinsing with distilled water). Burettes should be placed in the burette holder to drain—upsidedown, with the stopcock open (Make sure you initial your burette).

(11) Label all your glassware (on the glassware itself) as you use it and you should write the labels in your notebook so that you can keep track of what sample goes with what piece of data.

(12) Refer to guidelines about keeping a notebook and writing a report in the ACS format.

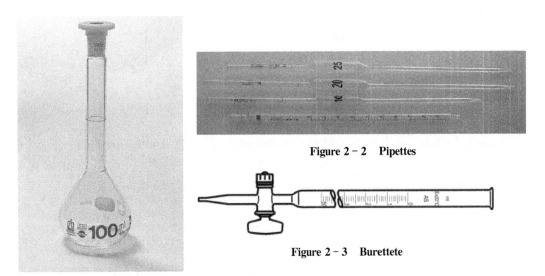

Figure 2 - 2   Pipettes

Figure 2 - 3   Burettete

Figure 2 - 1   Volumetric Flask

In precise work it is never safe to assume that the volume delivered by or contained in any volumetric instrument is exactly the amount indicated by the calibration mark. Instead, recalibration is usually performed by weighing the amount of water delivered by or contained in the volumetric apparatus. This mass is then converted to the desired volume using the tabulated density of Water:

$$\text{Volume} = \text{mass/density} \qquad (2\text{-}1\text{-}1)$$

All volumetric glassware should be either purchased with a Calibration Certificate or calibrated by the analyst in this manner.

II. Systematic errors affecting volumetric measurements

The volume occupied by a given mass of liquid varies with temperature, as does the volume of the device that holds the liquid. 20 ℃ has been chosen as the normal temperature for calibration of many volumetric glassware.

Glass is a fortunate choice for volumetric ware as it has a relatively small coefficient of thermal expansion; a glass vessel which holds 1. 000 00 L at 15 ℃ holds 1. 000 25 L at 25 ℃. If desired, the volume values (V) obtained at a temperature (t) can be corrected to 20 ℃ by use of:

$$V_{20} = V[1 + 0.000\ 025(20 - t)] \qquad (2\text{-}1\text{-}2)$$

Where V_{20} is the volume at 20 ℃, V is the volume value at a temperature t.

In most work, this correction is small enough it may be ignored.

However, the thermal expansion of the contained liquid is frequently of importance. Dilute aqueous solutions have a coefficient of thermal expansion of about 0. 025%/℃. A liter of water at 15 ℃ will occupy 1. 002 L at 25 ℃. A correction for this expansion must frequently be applied during calibration procedures.

Parallax is another source of error when using volumetric ware. A correction for this expansion must frequently be applied during calibration procedures. Frequently, graduation marks encircle the apparatus to aid in this.

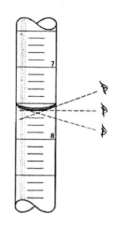

Fiigure 2 - 4　Parallax

Readings which are either too high or too low will result otherwise.

III. Tips forcorrect use of volumetric glassware

A. Pipettes

The Pipette is used to transfer a volume of solution from one container to another. Most volumetric pipettes are calibrated to-deliver (TD); with a certain amount of the liquid remaining in the tip and as a film along the inner barrel after delivery of the liquid. The liquid in the tip should not be blown-out. Pipettes of the "blow-out" variety will usually have a ground glass ring at the top. And, drainage rates from the pipette must be carefully controlled so as to leave a uniform and reproducible film along the inner glass surface. Measuring pipettes will be gradated in appropriate units.

Once the pipette is cleaned and ready to use, make sure the outside of the tip is dry. Then rinse the Pipette with the solution to be transferred. Insert the tip into the liquid to be used and draw enough liquid into the pipette to fill a small portion of the bulb. Hold the liquid in the bulb by placing your fore finger over the end of the stem.

Withdraw the pipette from the liquid and gently rotate it at an angle so as to wet all portions of the bulb. Drain out and discard the rinsing liquid. Repeat this once more.

Figure 2 – 5 Use of pipette

To fill the pipette, insert it vertically in the liquid, with the tip near the bottom of the container. Apply suction to draw the liquid above the graduation mark. Quickly place a fore finger over the end of the stem. Withdraw the pipette from the liquid and use a dry paper to wipe off the stem. Now place the tip of the pipette against the container from which the liquid has been withdrawn and drain the excess liquid such that the meniscus is at the graduation mark.

Move the pipette to the receiving container and allow the liquid to flow out (avoiding splashing) of the pipette freely. When most of the liquid has drained from the pipette, touch the tip to the wall of the container until the flow stops and for an additional count of 10.

B. Volumetric flasks

The volumetric flask is used to prepare standard solutions or in diluting a sample. Most of these flasks are calibrated to-contain (TC) a given volume of liquid. When using a flask, the solution or solid to be diluted is added and solvent is added until the flask is about two-thirds full. It is important to rinse down any solid or liquid which has adhered to the neck. Swirl the solution until it is thoroughly mixed. Now add solvent until the meniscus is at the calibration mark. If any droplets of solvent adhere to the neck, use a piece of tissue to blot them out. Stopper the flask securely and invert the flask at least 10 times.

C. Burettes

The Burette is used to accurately deliver a variable amount of liquid. Fill the burette to above the zero mark and open the stopcock to fill the tip. Work air bubbles out of the tip by rapidly squirting the liquid through the tip or tapping the tip while solution is draining.

The initial burette reading is taken a few seconds, ten to twenty, after the drainage of liquid has ceased. The meniscus can be highlighted by holding a white piece of paper with a heavy black mark on it behind the burette.

Place the flask into which the liquid is to be drained on a white piece of paper. (This is done during a titration to help visualize color changes which occur during the titration.) The flask is swirled with the right-hand while the stopcock is manipulated with the left-hand.

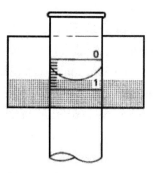

Figure 2 - 6　Use of burette

The burette should be opened and allowed to drain freely until near the point where liquid will no longer be added to the flask. Smaller additions are made as the end-point of the addition is neared. Allow a few seconds after closing the stopcock before making any readings. At the end-point, read the burette in a manner similar to that above.

As with pipettes, drainage rates must be controlled so as to provide a reproducible liquid film along the inner barrel of the Burette.

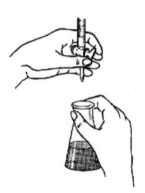

Figure 2 - 7

IV. Cleaning volumetric glassware

Cleaning of volumetric glassware is necessary to not only remove any contaminants, but also to ensure its accurate use. The film of water which adheres to the inner glass wall of a container as it is emptied must be uniform.

Two or three rinsings with tap water, a moderate amount of agitation with a dilute detergent solution, several rinsings with tap water, and two or three rinsings with distilled water are generally sufficient if the glassware is emptied and cleaned immediately after use.

If needed, use a warm detergent solution (60—70 ℃). A burette or test tube brush can be used in the cleaning of burettes and the neck of volumetric flasks. Volumetric flasks can be filled with cleaning solution directly. Pipettes and burettes should be filled by inverting them and drawing the cleaning solution into the device with suction. Avoid getting cleaning solution in the stopcock. Allow the warm cleaning solution to stand in the device for about 15 minutes; never longer than 20 minutes. Drain the cleaning solution and rinse thoroughly with tap water and finally 2—3 times with distilled water.

Pipettes and burettes should be rinsed at least once with the solution with which they are to be filled before use.

REAGENTS AND APPARATUS

- Analytical balance, weighing bottles, 400 mL beakers, 100 mL plastic bottles with caps, 5 mL and 25 mL volumetric pipettes, 25 mL volumetric flasks, 50 mL burettes, thermometer, crucible tong.
- Distilled and deionized water.

PROCEDURES

Part A—Use of the analytical balance

The analytical balances in the lab are probably the most precise, accurate and reliable pieces of equipment that you will use during the semester. Although there are inherent limits in the accuracy and precision of these balances most weighing errors are caused by incorrect handling of the sample. In the first part of this lab you will investigate several potential sources of error.

1. In this section you will determine the mass of a clean, dry weighing bottle under various conditions. Unless instructed otherwise, you should handle the bottle with your crucible tongs, gloves, or lint-free paper and measurements should be made to the nearest 0.1 mg. Begin by placing the weighing bottle and cap (with cap removed) in the oven for about 5 minutes. Remove and re-mass while warm. Follow the change in its apparent mass for several minutes, reweighing every thirty seconds. Record all masses including the final constant value.

2. After massing the weighing bottle, roll it around in your hand (handle the bottle with your fingers) and then re-mass and compare the two masses.

3. Next, wipe the bottle clean with a dry, lint-free cloth or tissue and reweigh. Record all observations.

4. Hold the weighing bottle and inch from your mouth and breathe on it several times. Re-mass and compare with previous data.

5. Discuss your results in your laboratory write-up.

Part B—Calibration of volumetric glassware

As was noted above, volumetric glassware is calibrated by measuring the mass of water that is contained in or delivered by the device.

To obtain an accurate mass measurement, buoyancy effects must be corrected for. The amount of air displaced by the standard weights of the balance is somewhat different than the amount of air displaced by the weighed water. This difference leads to different buoyancies for these objects; meaning the balance levels at a point other than when the two objects are of the same mass. This can be corrected for use:

$$m_{true} = m_{meas} + d_a((m_{meas}/d) - (m_{meas}/d_s))$$ (2-1-3)

Where d_s is the density of the standard weights (8.47 g/cm^3), d_a is the density of air (0.001 2 g • cm^3), and d is the density of the object being measured.

This mass data is then converted to volume data using the tabulated density of Water (See Appendix) at the temperature of calibration. (In very accurate work, the thermometer must also be calibrated as an incorrect temperature reading will lead to the use of an incorrect density for water. This, in turn, will give an inaccurate volume calibration.)

Finally, this volume data is corrected to the standard temperature of 20 ℃. This can be accomplished using the thermal expansion coefficient of water: 0.000 25/ ℃:

$$V_{20} = V[1 + 0.000\ 25(20 - t)]$$ (2-1-4)

Thus, in this exercise we will calibrate a volumetric flask and a pipette and determine a burette correction factor by calibrating each of these devices with water. In each case, the measured mass of the calibrating water will be corrected for buoyancy effects and the resulting volume will be standardized to 20 ℃.

1. Cleaning

Begin by cleaning a 5 mL or 25 mL volumetric pipette, a 50 mL burette, and a 25 mL volumetric flask according to the procedure outlined above. It is imperative for the purposes of calibration that these glassware items be cleaned such that water drains uniformly and does not leave breaks or droplets on the walls of the glass.

If detergent solutions are not sufficient to clean your glassware, a cleaning solution (dichromate in conc. sulfuric acid) may be used. Consult you instructor before taking this step.

Once cleaned, the burette should be filled with distilled water and clamped in an upright position and stored in this manner until needed. The volumetric flask should be clamped in an inverted position so that it may dry.

2. Calibration of a pipette

Use your cleaned Pipette. Weigh a receiving container on the analytical balance: a 100 mL plastic beaker with an aluminum foil cover. Pipette distilled water into the plastic beaker and reweigh it.

Record the temperature of the water used.

Repeat the procedure at least 2 more times. Dry the plastic beaker and re-weigh it for each replication. (Are you pipetteing consistently and correctly?)

Calculate the apparent mass and the buoyancy corrected mass of the water delivered for each time you pipette. From this mass, and the density of water at the given temperature (See Appendix), calculate the volume of the water delivered. Correct the volume to 20 ℃. Calculate the average, standard deviation, and 90%

confidence interval for your calibration result.

Is your result within the listed tolerance for this pipette? (See Appendix. What is the better question to ask?)

3. Calibration of a burette

Use your cleaned 50 mL burette. Fill the burette with water. Make sure the tip is free of bubbles. Drain into a waste beaker until it is at, or just below, the zero mark. Allow 10—20 seconds for drainage. Make an initial reading to a precision of at least 0.01 mL. Test for tightness of the stopcock by allowing the burette to stand for 5 minutes and then re-reading the volume. There should be no noticeable change in the reading.

Once the tightness of the stopcock is assured, refill the burette and again drain into a waste until it is at, or just below, the zero mark. Allow for drainage. Touch the tip of the burette to the wall of the waste beaker to remove the pendent drop of water. Make a volume reading.

Weigh a receiving container on the analytical balance; a 100 mL plastic beaker with aluminum foil cover. Drain about 5 mL of water from the burette into the beaker. Allow 10—20 seconds for drainage. Touch the tip of the burette to the wall of the beaker to remove the pendent drop again. Read the burette and weigh the water.

Calculate the actual volume of water delivered by the burette in the same manner as outlined above in the procedure on calibrating pipettes. Calculate the correction factor (CF) by subtracting the apparent volume delivered, as given by the burette readings, from the actual volume delivered.

Repeat the procedure at least once more. The two correction factors should agree within 0.04 mL. If they do not, repeat the procedure again. Report the average correction factor for 5 mL.

Repeat this process for 15 mL, 25 mL, 35 mL, and 45 mL delivered.

Plot the average burette correction factor vs. volume delivered using Excel or some other graphing software.

Label and store your burette properly; upright and filled with distilled water. This is the burette you will use for the remainder of the course.

4. Calibration of a volumetric flask

Use your cleaned 25mL volumetric flask. Weigh the flask empty. Fill the flask to the mark and re-weigh it. Measure the temperature of the water used.

Repeat the procedure at least twice.

Calculate the true volume of the flask using the method outlined above. Report the average, standard deviation, and 90% confidence interval for this result.

DATA TREATMENT

Part A—Use of the analytical balance

1. Temperature effect

	30s	60s	90s	120s	150s	180s	210s	240s	270s	300s
mass(g)										

2. Under different conditions

	initial	after roll	after wipe	after breath
mass (g)				

Part B—Calibration of volumetric glassware

1. Calibration of apipette (V_{read} = ___ mL, temperature= ___ ℃)

	1	2	3
Container$_{meas}$ (g)			
container+H_2O_{meas} (g)			
H_2O_{meas} (g)			
H_2O_{true} (g)			
$V(pipette)_{true}$ (mL)			
$\Delta V(V_{true}-V_{read})$ (μL)			
$\Delta \bar{V}(\mu L)$			
σ			
90% confidence interval			

2. Calibration of a burette (temperature= ___ ℃)

	5 mL	15mL	25 mL	35 mL	45 mL
$V1_{initial}$ (mL)					
$V1_{end}$ (mL)					
$m1_{beaker}$ (g)					
$m1_{bearker+H_2O}$ (g)					
H_2O1_{meas} (g)					
H_2O1_{true} (g)					

(**Continued**)

	5 mL	15mL	25 mL	35 mL	45 mL
$V(\text{burette})1_{\text{true}}$ (mL)					
$CF(\Delta V1)$ (μL)					
$V2_{\text{initial}}$ (mL)					
$V2_{\text{end}}$ (mL)					
$m2_{\text{beaker}}$ (g)					
$m2_{\text{bearker}+H_2O}$ (g)					
H_2O2_{meas} (g)					
H_2O2_{true} (g)					
$V(\text{burette})2_{\text{true}}$ (mL)					
$CF(\Delta V2)$ (μL)					
$\overline{CF(\Delta \bar{V})}$					

3. Plot the average $\overline{CF}$ of burette vs. Volume.

4. Calibration of a flask (V_{read}＝ mL, temperature＝ ℃)

	1	2	3
$\text{beaker}_{\text{meas}}$ (g)			
$\text{beaker}+H_2O_{\text{meas}}$ (g)			
H_2O_{meas} (g)			
H_2O_{true} (g)			
$V(\text{flask})_{\text{true}}$ (mL)			
$\Delta V(V_{\text{true}}-V_{\text{read}})$ (μL)			
$\Delta \bar{V}(\mu L)$			
σ			
90% confidence interval			

QUESTIONS

1. How should weighing bottles be handled on a regular basis?

2. Are your results within the listed tolerance for pipettes, burettes, and flasks? (See Appendix.) Analysis the reasons.

Appendix

Table 2 - 1   Density of Water

Temperature(℃)	Density(g/mL)
10	0. 999 702 6
11	0. 999 608 4
12	0. 999 500 4
13	0. 999 380 1
14	0. 999 247 4
15	0. 999 102 6
16	0. 998 946 0
17	0. 998 777 9
18	0. 998 598 6
19	0. 998 408 2
20	0. 998 207 1
21	0. 997 995 5
22	0. 997 773 5
23	0. 997 541 5
24	0. 997 299 5
25	0. 997 047 9
26	0. 996 786 7
27	0. 996 516 2
28	0. 996 236 5
29	0. 995 947 8
30	0. 995 650 2

Table 2 - 2   olerances for Class A Volumetric Glassware at 20 ℃

Pipettes		Volumetric Flasks	
Capacity(mL)	Tolerances (mL)	Capacity(mL)	Tolerances (mL)
0. 5	0. 006	5	0. 02
1	0. 006	10	0. 02
2	0. 006	25	0. 03
5	0. 01	50	0. 05
10	0. 02	100	0. 08
20	0. 03	250	0. 12
25	0. 03	500	0. 20
50	0. 05	1 000	0. 30
100	0. 08		

Burettes

Capacity(mL)	Tolerances (mL)
5	0. 01
10	0. 02
25	0. 03
50	0. 05
100	0. 20

2.2　Practice on Acid-base Titration

AIMS

　　1. To understand the concept of titrimetric analysis.

　　2. To learn how to prepare hydrochloric acid solution and sodium hydroxide solution.

　　3. Practice on the basic operation of titration.

INTRODUCTION

　　A "titration" is a common laboratory method of quantitative chemical analysis that is used to determine the unknown concentration of a known analyte. Titrations belong to a class of analytical techniques known as "volumetric analysis". Since volumes can be precisely delivered and measured using standard laboratory equipment, titration techniques can yield both accurate and precise measurements if care is taken by the analyst. Accuracy is defined as the closeness of a result (usually the average of several measurements) to a known accepted value. Precision relates to the closeness of the measurements themselves. In any analytical experiment, one strives for both accuracy and precision to validate the results.

　　In this experiment you will be graded on both your accuracy (how close you are to the actual value) and the precision of your results (how reproducible they are).

　　In a titration an analyst titrates a solution of unknown concentration with a solution of known concentration (a standard solution). The stoichiometry of the reaction between the standard and the analyte is known.

　　Using a calibrated "burette" to add the standard solution, it is possible to determine with accuracy the amount of analyte present in the unknown solution when the titration endpoint is reached. The endpoint of a titration is the point at which the titration is complete. The endpoint is generally signaled by an indicator (see below) that is added to the analyte solution. This is ideally the same volume or very close to as the equivalence point—the volume of added titrant at which the number of moles of titrant is equal to the number of moles of analyte, or some multiple thereof. In the classic strong acid-strong base titration, the endpoint of a titration is the point at which the pH of the reactant is just about equal to 7, and often when the solution takes on a persisting solid color as in the pink of phenolphthalein indicator used in this experiment.

At the equivalence of a titration, the moles of the analyte are calculated which mayyield quantitative information such as concentration (mol • L^{-1}) or molar mass (g • mol^{-1}).

The ability to perform a titration experiment well is a crucial skill that all general chemistry students must master. You will be conducting a titration for the next lab and, since most of you have yet to perform such experiments, we are offering you a chance to practice beforehand. Recall, as they say, "practice makes perfect."

The goal of this experiment will be to practice and perfect your volumetric techniques using pipettes and a burette. Your lab instructor will demonstrate proper pipet and titration techniques.

Example: the titration of 20 mL of a 0. 100 0 mol • L^{-1} strong monoprotic acid HCl with 20 mL of a 0. 100 0 mol • L^{-1} strong base NaOH.

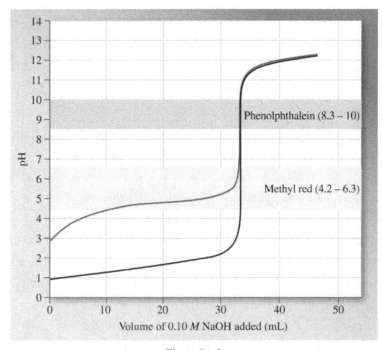

Figure 2 - 8

Reaction: $HCl + NaOH \longrightarrow NaCl + H_2O$, so the moles of HCl is equal to the moles of NaOH, $(cV)_{HCl} = n(HCl) = n(NaOH) = (cV)_{NaOH}$

$$\frac{c(HCl)}{c(NaOH)} = \frac{V(NaOH)}{V(HCl)} \tag{2-2-1}$$

REAGENTS AND APPARATUS

- Analytical balance, 250 mL volumetric flasks, 50 mL burettes, 250 mL conical beakers, 25 mL volumetric pipettes, 25 mL beakers, 25 mL volumetric pipettes, droppers, glass rods, rubber bulbs
- Concentrated hydrochloric acid (HCl), sodium hydroxide (NaOH), indicator: 0.1% methyl orange, 0.1% phenolphthalein solution, distilled and deionized water

PROCEDURES

1. Preparation of 0.1 mol · L^{-1} HCl solution

Using a volumetric pipet and a bulb, pipet 4.2 mL of about 6 M concentrated HCl(aq) into a 250 mL volumetric flask. (Record the exact concentration (mol/L) written on the container).

$$4.2 \text{ mL HCl} \xrightarrow[\text{dilute fill to the mark}]{} 250 \text{ mL}$$

2. Preparation of 0.1 mol · L^{-1} NaOH solution

Weigh about 1.0 g NaOH, dissolve it completely in beaker, then transfer into a 250 mL volumetric flask.

$$1.0 \text{ g NaOH} \xrightarrow[\text{dissolve dilute fill to the mark}]{} 250 \text{ mL}$$

3. Titration practice

3.1 Titrate 0.1 mol · L^{-1} NaOH with 0.1 mol · L^{-1} HCl (indicator: MO)

Retrieve a burette from the wooden case in your lab. Inspect it carefully to see that the glass end to the stopcock is not plugged and that the stopcock does not leak. Empty the burette into the sink (save the stopper) and rinse your burette with a small amount of deionized water (~10 mL) three times.

Close the stopcock and rinse your burette with a small amount of the prepared HCl solution (~5 mL) three times. On the third rinse, allow the acid to drain from the stopcock. You can use one of your large beakers to collect waste. Empty the waste beaker into the sink at the end of the experiment.

Fill your burette with the prepared HCl solution, making sure that there are no air bubbles in the stopcock tip. (Make sure that the stopcock is closed first!)

With a 25.00 mL pipet and a bulb, pipet the prepared NaOH solution to a conical beaker. Add 2 or 3 drops of methyl orange indicator to the conical beaker.

Record the initial burette reading on the data sheet provided. All volume recordings should be to ±0.01 mL! Add HCl drop by drop while swirling the conical beaker at the same time. When the color changes from yellow to red, you have reached the titration endpoint. Record the final burette reading (±0.01 mL).

3.2   Titrate $0.1 \, mol \cdot L^{-1}$ HCl with $0.1 \, mol \cdot L^{-1}$ NaOH (indicator: PP)

Clean the burette thoroughly, and then rinse your burette with a small amount of the prepared NaOH solution (~ 5 mL) three times. Fill your burette with the prepared NaOH solution, making sure that there are no air bubbles in the stopcock tip.

Pipet 25.00 mL NaOH solution to a conical beaker. Add 2 or 3 drops of phenolphthalein indicator to the beaker.

Record the initial burette reading on the data sheet provided. Add NaOH drop by drop while swirling the conical beaker at the same time. When the faint pink persists for 30 sencond, you have reached the titration endpoint. you have reached the titration endpoint. Record the final burette reading. All volume recordings should be to ±0.01 mL!

DATA TREATMENT

1. Data for titration of $0.1 \, mol \cdot L^{-1}$ NaOH with $0.1 \, mol \cdot L^{-1}$ HCl

		1	2	3
	NaOH analyte/mL	25.00	25.00	25.00
HCl titrant	before titration/mL			
	after titration/mL			
	net volume/mL			
Data report	volume ratio $\dfrac{V(\text{NaOH})}{V(\text{HCl})}$			
	average volume ratio			
	the average relative deviation			

2. Data fortitration of 0. 1 mol · L⁻¹ HCl with 0. 1 mol · L⁻¹ NaOH

		1	2	3
HCl analyte/mL		25. 00	25. 00	25. 00
NaOH titrant	before titration/mL			
	after titration/mL			
	net volume/mL			
Data report	volume ratio $\dfrac{V(\text{NaOH})}{V(\text{HCl})}$			
	average volume ratio			
	the average relative deviation			

QUESTIONS

1. Can we prepare HCl or NaOH standard solution with accurate concentration directly?

2. What is meant by accuracy and precision in a scientific measurement?

3. In this experiment, is the volume ratio of titration NaOH with HCl and titration HCl with NaOH the same?

2. 3 Determination of Nitrogen Content in Ammonium Salt (Formaldehy de Method)

AIMS

1. To learn to determine the solution concentration with the primary standard solution.

2. To determine the nitrogen content in ammonium salt with formaldehyde.

INTRODUCTION

Common ammonium salts, such as ammonium sulfate, ammonium chloride, ammonium nitrate, are strong acid and weak alkali salts.

$$NH_4^+ + H_2O \rightleftharpoons NH_3 \cdot H_2O + H^+$$

$$K_a(NH_4^+) = K_w/K_b(NH_3 \cdot H_2O) = 5.6 \times 10^{-10}$$

Although NH_4^+ is acidic, it can not be titrated directly because of $cK_a < 10^{-8}$, where c is the concentration, K_a is the acid dissociation constant, K_w is the ion product of water, K_b is the base dissociation constant.

There are three more or less standard methods in general use for analyzing ammonium nitrate, i. e., the Kjeldahl method, the nitrometer method, and the impurity method. The first two are very accurate, but take from ¾ to 1 ½ hrs for duplicate analyses, which is too long for most control work. The third method, while commonly used, is unsatisfactory, in that any errors of omission or commission are combined here.

The method in our experiment, making use of formaldehyde, is much more rapid and quite accurate. The method depends upon the ease with which formaldehyde reacts with ammonia or ammonium salts to form hexamethylenetetramine:

$$4NH_4^+ + 6\ HCHO = (CH_2)_6N_4H^+ + 3\ H^+ + 6\ H_2O$$

$$K_a[(CH_2)_6N_4H^+] = 7.1 \times 10^{-6}$$

In the circumstance, $cK_a[(CH_2)_6N_4H^+] > 10^{-8}$, meet the requirements of direct titration.

The method is very simple and in general is as follows: A neutral 20 percent solution of formaldehyde is added to the ammonium nitrate solution, and titrated with standard NaOH, using phenolphthalein as indicator, since the latter is not sensitive to hexamethylenetetramine. The amount of NaOH used gives the amount of ammonium nitrate present.

$$(CH_2)_6N_4H^+ + 3H^+ + 4OH^- = (CH_2)_6N_4 + 4H_2O$$

Mole relations: $n(N) = n(NH_4^+) = n(H^+) = (cV)_{NaOH}$

Nitrogen content: $N\% = \dfrac{n(N) \times 14.01}{\text{sample mass}} = \dfrac{(cV)_{NaOH} \times 14.01}{\text{sample mass} \times 10^3}$

REAGENTS AND APPARATUS

- Analytical balance, 100 mL and 250 mL volumetric flasks, 50 mL burettes, 250mL conical beakers, 25 mL beaker, 10 mL graduated cylinder, droppers, glass rods, rubber bulbs
- Sodium hydroxide (NaOH), potassium hydrogen phthalate (KHP), ammonium sulphate ($(NH_4)_2SO_4$), 40% formaldehyde solution, 10 g $\cdot$ L^{-1} phenolphthalein (PP) solution, distilled and deionized water

PROCEDURES

1. Treatment of neutral formaldehyde solution

Formaldehyde often contains trace formic acid, which is caused by the oxidation by air. Formic acid should be removed, or otherwise it will bring positive error to the total reaction. Take the supernatant of the 40% formaldehyde in beaker, dilute it twice with water. Add 1—2 drops of phenolphthalein solution as indicator, and then titrate with 0.1 mol · L^{-1} NaOH standard solution until the solution is light pink, finally drop with unneutralized formaldehyde to just colorless.

2. Preparation and standardization of 0.1 mol · L^{-1} NaOH solution

About 1.0 g sodium hydroxide is weighed (record the exact weight on your paper, and the weighing accuracy should be to ±0.000 1 g), and dissolved in water in a small beaker (25 mL). After all dissolved, transfer to 250 mL volumetric flasks and add water to the calibration line.

$$1.0 \text{ g NaOH} \xrightarrow{\text{H}_2\text{O dissolve}} \xrightarrow{\text{transfer}} \xrightarrow{\text{fill to the mark}} 250 \text{ mL}$$

Standardization (3 times)

Accurately weigh 0.4—0.5 g potassium hydrogen phthalate (the weighing accuracy should be to ±0.000 1 g), dissolve in a small beaker (25 mL) and then transfer to a conical beaker. The beaker needs to be rinsed three times with distilled water, and all the solution after each rinse should be transferred to the conical beaker. The total amount of water is 50 mL, including dissolution and rinsing.

Add 1—2 drops of phenolphthalein solution to the conical beaker as indicator, and record the initial burette reading, and then titrate with 0.1 mol · L^{-1} NaOH solution slowly while swirl the conical beaker. When the color turns to light pink, and does not fade for 30 seconds, the end point is reached. Record the final burette reading. The burette accuracy should be 0.01 mL. Repeated three times.

$$0.4{\sim}0.5 \text{ g KHP} \xrightarrow{\text{50 mL H}_2\text{O}} \xrightarrow{\text{1D PP}} \xrightarrow{\text{NaOH titrate}} \text{pink solution}$$

The reaction for potassium hydrogen phthalate with sodium hydroxide:

$$\text{KHP} + \text{NaOH} =\!=\!= \text{KNaP} + \text{H}_2\text{O}$$

Mole relations: $\dfrac{m(\text{KHP}) \times 10^3}{204.1} = n(\text{KHP}) = n(\text{NaOH}) = (cV)_{\text{NaOH}}$

3. Determine the nitrogen content of the sample

3.1 Preparation sample solution

Accurately weigh 0. 6—0. 8 g $(NH_4)_2SO_4$ in beaker (the weighing accuracy should be to ± 0.0001 g), dissolve it with distilled water, then move it quantitatively to 100 mL volumetric flask, and finally dilute it with distilled water to the mark and shake it evenly.

$$0.6 \sim 0.8 \text{ g } (NH_4)_2SO_4 \xrightarrow[\text{transfer} \quad \text{fill to the mark}]{\text{H}_2\text{O dissolve}} 100 \text{ mL}$$

3.2 Titraion with NaOH

Transfer 25. 00 mL of sample solution into a conical beaker with pipette. Add 4 mL neutral formaldehyde solution, shake well and let stand for 1 minute. 1 drop of phenolphthalein indicator is added to the solution, and then titrated with the calculated NaOH solution until the solution is light pink for 30 seconds. Record the initial and final burette reading with accuracy of 0. 01 mL. The endpoint should be determined for three times in parallel.

$$25 \text{ mL smaple} \xrightarrow[\text{shake up} \quad \text{static for 1 min} \quad \text{NaOH titrate}]{4 \text{ mL 20\%HCHO} \quad \text{1D PP}} \text{pink solution}$$

DATA TREATMENT

1. Data for standardization of 0. 1 mol · L^{-1} NaOH solution

times		1	2	3
KHP weighting	Mass m(KHP)/g			
Titration with NaOH	initial reading/mL			
	final reading/mL			
	net volume/mL			
Results	c(NaOH)/mol · L^{-1}			
	$\bar{c}$(NaOH)/mol · L^{-1}			
	the average relative deviation %			

2. Data for determination of the sample's nitrogen content

times		1	2	3
$(NH_4)_2SO_4$ weighting	$m[(NH_4)_2SO_4]/g$			
$(NH_4)_2SO_4$ solution/mL		25.00	25.00	25.00
Titration with NaOH	initial reading/mL			
	final reading/mL			
	net volume/mL			
Results	nitrogen content %			
	Average nitrogen content %			

QUESTIONS

1. If there is little amount of formicacid in the formaldehyde sample, what effects does it have? How to eliminate its influence?

2. Except potassium hydrogen phthalate(KHP) primary standard, what else can be used to standardize the concentration of NaOH solution?

3. In the titration of NaOH solution with potassium hydrogen phthalate (KHP), why is weighting the mass of KHP for 0.4~0.5 g?

2.4 Determination of the Hardness of Water

AIMS

1. To master the principles, methods and calculation of complexometric titration.

2. To know how to use indicators and their color change.

INTRODUCTION

What is hard water, and why does it have this effect? Hard water contains a higher than normal concentration of calcium and magnesium ions, since most hardness is caused by carbonate mineral deposits. These ions form precipitates with soap, causing the build up of soap scum. Additionally, since the soap molecules are being precipitated by the Ca^{2+} and Mg^{2+} ions, there is less soap available to form lather.

Taking a shower in hard water can be very frustrating! These cations form insoluble salts with a reagent in soap, decreasing its cleaning effectiveness.

Another effect of hard water is "boiler scale". When hard water comes into contact with dissolved carbonates, a precipitate of insoluble calcium carbonate can form. This "scale" can build up on the inside of water pipes to such a degree that the pipes become almost completely blocked. The following chart shows how hard water is classified. For reporting purposes, hardness is reported as parts per million (ppm) of $CaCO_3$ (by weight). A water supply with a hardness of 100 ppm contains the equivalent of 100 g of $CaCO_3$ in 1 million g of water or 0.1 g in 1 L of water (or 1 000 g of water since the density of water is about 1 g/mL). In other words, even though both Ca^{2+} and Mg^{2+} contribute to water hardness, it is reported as though all hardness ions are Ca^{2+} from $CaCO_3$. Since Ca^{2+} and Mg^{2+} behave exactly the same, this convention is convenient shorthand.

Hardness (ppm $CaCO_3$)	Classification
15 ppm	Very Soft
15 ppm—50 ppm	Soft
50 ppm—100 ppm	Medium hard
100 ppm—200 ppm	Hard
> 200 ppm	Very hard

The hardness of a sample of water can be measured by determining the concentration of the dissolved Ca^{2+} and Mg^{2+} ions. The procedure you will use is called a titration. To analyze for Ca^{2+} and Mg^{2+} ions you will add a substance, Na_2EDTA the chelating agent, which will react with the metal ions and remove them from solution.

$$Ca^{2+} + EDTA^{2-} = CaEDTA$$
$$Mg^{2+} + EDTA^{2-} = MgEDTA$$

Na_2EDTA is a complex molecule. Its name stands for ethylenediaminetetraacetic acid-disodium salt. The formula of disodium EDTA is $Na_2C_{10}H_{14}N_2O_8$ with the formula mass of 372.24 g·mol^{-1}.

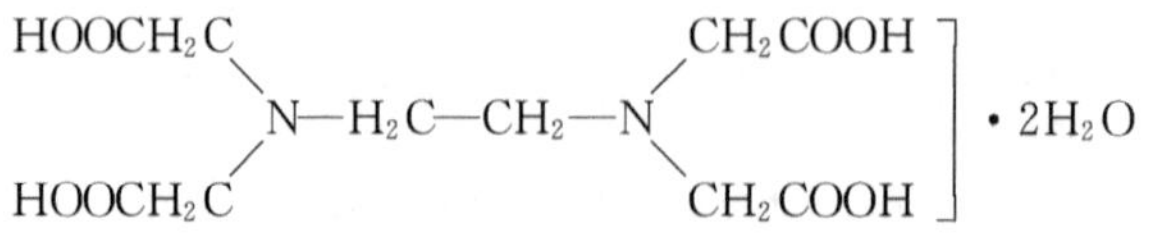

Figure 2-9   Structure of disodium EDTA dehydrate (372.24 g·mol^{-1})

The four acid oxygen sites and the two nitrogen atoms have unshared electron pairs, which can form bonds to a metal ion forming a complex ion or coordination compound. The complex is quite stable, and the conditions of its formation can ordinarily be controlled so that it is selective for a particular metal ion. EDTA reacts with Ca^{2+} and Mg^{2+} in a one to one mole ratio.

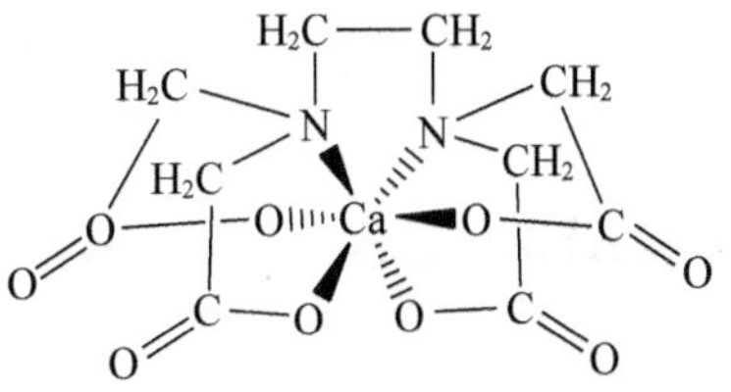

Figure 2 − 10   Structure of CaEDTA

In a titration to determine the concentration of a metal ion, the added EDTA combines quantitatively with the cation to form the complex. The endpoint occurs when essentially all of the cation has reacted.

Since not only EDTA but also Ca^{2+} and Mg^{2+} are colorless, it is necessary to use a special indicator to detect the endpoint of the titration. The indicator most often used is called eriochrome black T, which forms a very stable wine-red complex, $MgIn^-$, with the magnesium ion. A tiny amount of this complex will be present in the solution during the titration. As EDTA is added, it will complex free Ca^{2+} and Mg^{2+} ions, leaving the $MgIn^-$ complex alone until essentially all of the calcium and magnesium have been converted to chelates. At this point EDTA concentration will increase sufficiently to displace Mg^{2+} from the indicator complex; the indicator reverts to its uncombined form, which is sky blue, establishing the endpoint of the titration.

The titration is carried out at a pH of 10, in a NH_3/NH_4^+ buffer, which keeps the EDTA mainly in the form HY^{3-}, where it complexes the Group 2 ions very well but does not tend to react as readily with other cations such as Fe^{3+} that might be present as impurities in the water. The equations for the reactions which occur during the titration are:

Titration reaction:

$$HY^{3-}(aq)+Ca^{2+}(aq)\!=\!\!=\!\!CaY^{2-}(aq)+H^+(aq) \text{ (also for } Mg^{2+})$$

Endpoint reaction:

$$HY^{3-}(aq)+MgIn^-(aq)\!=\!\!=\!\!MgY^{2-}(aq)+HIn^{2-}(aq)$$

$$\qquad\qquad\ \text{wine red} \qquad\quad \text{sky blue}$$

Since the indicator requires a trace of Mg^{2+} to operate properly, it is suitable for determination of the total content of Ca^{2+} and Mg^{2+} ions. If Ca^{2+} ions are measured alone, you need to change the indicator to calmagite due to the insensitive response of eriochrome black without Mg^{2+}.

At this case, the titration is carried out at a pH of 12, in a NaOH buffer, toprevent the interference of Mg^{2+} ions.

Titration reaction: $HY^{3-}(aq) + Ca^{2+}(aq) = CaY^{2-}(aq) + H^+(aq)$

End point reaction: $HY^{3-}(aq) + CaIn^-(aq) = CaY^{2-}(aq) + HIn^{2-}(aq)$
$$\qquad\qquad\qquad\text{wine red}\qquad\text{sky blue}$$

REAGENTS AND APPARATUS

- 50 mL burettes, 250 mL conical beakers, 50 mLvolumetric pipettes, 10 mL graduated cylinder, droppers, rubber bulbs.
- Water sample, 0. 02 mol · L^{-1} disodium ethylenediamine tetraacetic (Na_2EDTA) solution, 0. 1 mol · L^{-1} NH_3/NH_4^+ buffer (pH = 10), 1 mol · L^{-1} sodium hydroxide (NaOH) solution (pH = 12), 5 g · L^{-1} Eriochrome black T solution, 5 g · L^{-1} calmagite solution, distilled and deionized water

PROCEDURES

1. Determine the total concentration of Ca^{2+} and Mg^{2+} ions

Wash a 50. 00 mL burette thoroughly with water, and then rinse it with distilled water and a few mLs of Na_2EDTA solution sequently. Drain through the stopcock and then fill the burette with the Na_2EDTA solution.

Pipet three 50. 00 mL water sample into three clean conical beakers. To each beaker, add 5 mL NH_3/NH_4^+ buffer to control the pH to about 10. Add 3—4 drops of eriochrome black T solution as indicator, and you will find the solution turns into wine red. Read the burette to 0. 01 mL and add Na_2EDTA to the solution until the last tinge of purple just disappears. The color change is rather slow, so titrate slowly near the endpoint. Read the burette to 0. 01 mL again to determine the volume needed.

Refill the burette, read it, and titrate the second solution, then the third.

2. Determine the concentration of Ca^{2+} ions

Refill the burette, Record to the nearest 0. 01 mL the exact liquid level of Na_2EDTA in the burette.

Pipet three 50. 00 mL water sample into three clean conical beakers. To each beaker, add 2 mL NaOH solution to control the pH to about 12. Add 3—4 drops of calmagite solution as indicator, and you will find the solution turns into wine red. Slowly add the Na_2EDTA solution from the burette until the endpoint is reached. The endpoint is the point at which the solution in the conical beakers turns a distinct blue.

Read the burette to 0. 01 mL again and calculate the volume used in the titration by subtracting the initial volume from the final volume of $Na_2 EDTA$.

Refill the burette, read it, and titrate the second solution, then the third.

DATA TREATMENT

1. Data for the total concentration of Ca^{2+} and Mg^{2+} ions

		1	2	3
	Water sample/mL	50. 00	50. 00	50. 00
Na_2 EDTA titrant	initial reading/mL			
	final reading/mL			
	net volume/mL			
	average volume $\bar{V}_1$/mL			

2. Data for the concentration of Ca^{2+} ions

		1	2	3
	Water sample/mL	50. 00	50. 00	50. 00
Na_2 EDTA titrant	initial reading/mL			
	final reading/mL			
	net volume/mL			
	average volume $\bar{V}_2$/mL			

3. Calculation of the water sample's hardness

You will use V_1, V_2, and $c(Na_2 EDTA) = 0. 02$ mol $\cdot$ L^{-1}

$$CaO \text{ content/mg} \cdot L^{-1} = \frac{c\,\bar{V}_1 \times M(CaO) \times 1\,000}{50.\,00}$$

Total permanent hardness = German degrees $^\circ dH$ (1 $^\circ dH$ = 10 mg CaO $\cdot$ L^{-1})

$$Ca^{2+} \text{ content/mg} \cdot L^{-1} = \frac{c\,\bar{V}_2 \times M(Ca) \times 1\,000}{50.\,00}$$

$$Mg^{2+} \text{ content/mg} \cdot L^{-1} = \frac{c(\bar{V}_1 - \bar{V}_2) \times M(Mg) \times 1\,000}{50.\,00}$$

QUESTIONS

1. Why do you need to select a pH range incomplexometric titrations?

2. What can be primary standard to standardize the Na_2 EDTA concentration?

3. Water is usually softened by using an ion exchange resin to replace each Ca^{2+} (and Mg^{2+}) ion with 2 Na^+ ions. What must be true of the Na^+ salt of soap?

2.5 Preparation and Calibration of Potassium Permangante Solution

INTRODUCTION

Permanganometry is one of the techniques used in qualitative analysis in Chemistry. It is a redox titration and involves the use of permanganates and is used to estimate the amount of analyte present in unknown chemical samples. It involves two steps, namely the titration of the analyte with potassium permanganate solution and then the standardization of potassium permanganate solution with standard sodium oxalate solution. The titration involves volumetric manipulations to prepare the analyte solution.

Potassium permanganate is an oxidizing agent. It can retain its concentration over a long period under proper storage conditions. The ability of potassium permanganate solution to oxidize is due to the conversion of MnO_4^- ion to Mn^{2+} in acidic solution, to MnO_4^{2-} in alkaline, and to MnO_2 in neutral solution. The MnO_4^- ion is reduced in accordance with the following reactions.

$$MnO_4^- + 8H^+ + 5e^- = Mn^{2+} + 4H_2O \text{ (acidic medium)}$$
$$MnO_4^- + 2H_2O + 3e^- = MnO_2 + 4OH^- \text{ (neutral medium)}$$
$$MnO_4^- + e^- = MnO_4^{2-} \text{ (alkaline medium)}$$

The reduction of permanganate requires strong acidic conditions. Reduction of purple permanganate ion to the colorless Mn^{2+} ion, the solution will turn from dark purple to a faint pink color at the equivalence point. The permanganate can act as self-indicator, and no additional indicator is needed. The reactions of permanganate in solution are rapid.

Potassium permanganate is not a primary standard, which contains a small amount of MnO_2 and other impurities, such as sulphate, chloride and nitrate, etc. In addition, distilled water often contains a small amount of organic substances, which can reduce potassium permanganate. So, the concentration of $KMnO_4$ is easy to change, and its' accurate concentration can not be calculated by data from direct preparation. If it is used for a long time, it must be calibrated regularly. The actual concentration can be standardized by using sodium oxalate or oxalic acid. The former is preferred over oxalic acid as available in a higher standard of purity (99.95%). It's available in the anhydrous form, with stability and no hygroscopicity.

$$2KMnO_4 + 5Na_2C_2O_4 + 8H_2SO_4 = K_2SO_4 + 2MnSO_4$$

Hence, based on the above theory our aim is to prepare and standardize potassium permanganate solution with sodium oxalate.

Points of attention:

1. Distilled water often contains a small amount of reducing substances, so that $KMnO_4$ can be reduced to MnO_2. The commercial $KMnO_4$ contains fine powdered $MnO_2 \cdot nH_2O$ as well. The reduced product MnO_2 can also act as a catalyst to accelerate the decomposition of $KMnO_4$. Therefore, the solution of $KMnO_4$ is usually boiled for a period of time. After cooling, it needs to be placed for 2 to 3 days to make it react adequately, and then the precipitate is filtered out.

2. At room temperature, the reaction rate between $KMnO_4$ and $Na_2C_2O_4$ is slow, and the reaction rate can be increased by heating. However, the temperature can not be too high. If the temperature exceeds 85 ℃, some $H_2C_2O_4$ will be decomposed. The reaction formula is as follows:

$$H_2C_2O_4 = CO_2 \uparrow + CO + H_2O$$

3. The acidity of sodium oxalate solution is about 1 mol · L^{-1} at the beginning of titration and 0.5 mol · L^{-1} at the end of titration, which can ensure the reaction to proceed normally and prevent the formation of MnO_2. If the brown turbidity (MnO_2) occurs in the titration process, H_2SO_4 should be added immediately to let the brown turbidity disappear.

4. At the beginning of titration, the reaction is very slow. Before the first drop of $KMnO_4$ fades completely, the second drop should not be added. The titration rate can be appropriately accelerated when Mn^{2+} is generated. However, if you titrate too fast, the local concentration of $KMnO_4$ is too high, and it will be decomposed, releasing oxygen or causing the oxidation of impurities, which will bring errors.

If the titration rate is too fast, some $KMnO_4$ will not react with $Na_2C_2O_4$ in the future, but will decompose in the following way:

$$4MnO_4^- + 4H^+ = 4MnO_2 + 3O_2 \uparrow + 2H_2O$$

5. The endpoint is unstable. When the solution appears pink and does not fade in

30 seconds, the titration can be considered completed. If there is any doubt about the endpoint, you can first record the reading and add a drop of $KMnO_4$ standard solution to confirm that the endpoint has arrived. The titration should not exceed the stoichiometric point.

6. Because the color of $KMnO_4$ solution is very deep, the meniscus of the liquid surface is not easy to see, so the reading should be from the highest edge of the liquid surface.

REAGENTS AND APPARATUS

- Analytical balance, heating mantles, 50 mL burettes, 250 mL conical beakers, 600 mL beakers, 25 mL graduated cylinders, droppers, glass rods, thermometers
- Potassium permanganate ($KMnO_4$), sodium oxalate ($Na_2C_2O_4$), 3 mol $\cdot$ L^{-1} sulphuric acid (H_2SO_4), distilled and deionized water

PROCEDURES

1. Preparation of $KMnO_4$ solution

Weighing 0. 8—0. 9 g solid $KMnO_4$, record the reading (the weighing accuracy should be to $\pm 0.000\,1$ g). Put it in a large beaker (like 600 mL), add 250 mL water. Because of the loss of water evaporation during boiling, appropriately 50 ml more water can be added. The potassium permanganate solution should be boiled for about 1 hour, then cool to room temperature, and filtered with a microporous glass funnel or glass wool funnel. Fill the filtrate in a brown flask and label it.

$$0.8\sim0.9 \text{ KMnO}_4 \xrightarrow[\text{dissolve}]{250 \text{ mL H}_2\text{O}} \xrightarrow[\triangle]{\text{boiling 1 h}} \xrightarrow{\text{cool}} \xrightarrow{\text{filtrate}} \text{filtrate (put aside)}$$

2. Standardization of $KMnO_4$ solution

Wash a 50. 00 mL burette thoroughly with water, and then rinse it with distilled water and a few mLs of the prepared $KMnO_4$ solution sequently. Drain through the stopcock and then fill the burette with the $KMnO_4$ solution.

Accurately weigh 0. 15—0. 20 g $Na_2C_2O_4$ three times (the weighing accuracy should be to $\pm 0.000\,1$ g). Place them in three 250 mL conical beakers, add about 30 mL water and 10 mL 3 mol $\cdot$ L^{-1} H_2SO_4 for each conical beaker. Heat the conical beaker slowly in a heating mantle, with a surface dish covered. When the temperature of $Na_2C_2O_4$ solution is in the range of 75—85 ℃ (the temperature at the beginning of steaming), just begin the titration with $KMnO_4$. Pay attention to your operation, it's not easy to swirl the conical beaker while heating.

The reaction speed needs to be slow at the beginning of titration. When Mn^{2+} is produced in solution, the titration speed can be properly accelerated. You'll find the pink disappear more and more quickly. When the solution is pink and lasts for 30 seconds without fading, the endpoint is reached. The concentration of $KMnO_4$ can be calculated according to the quality of $Na_2C_2O_4$ and the volume of $KMnO_4$ solution consumed.

Recycle the $KMnO_4$ solution after the experiment.

$$0.15 \sim 0.20 \text{ g } Na_2C_2O_4 \xrightarrow[\text{3 mol/L } H_2SO_4]{\text{40 mL } H_2O} \xrightarrow{\text{10 mL}} \xrightarrow[\triangle]{75 \sim 85 \text{ ℃}} \xrightarrow[\text{titrate}]{KMnO_4} \text{pink color}$$

(end point)

DATA TREATMENT

		1	2	3
$Na_2C_2O_4$ weighting	$m(Na_2C_2O_4)/g$			
$KMnO_4$ titrant	before titration/mL			
	after titration/mL			
	net volume/mL			
$c(KMnO_4)/\text{mol} \cdot L^{-1}$				
average $\bar{c}/\text{mol} \cdot L^{-1}$				
the average relative deviation				

QUESTIONS

1. Why can not one prepare the $KMnO_4$ standard solution directly?

2. When standardizing the concentration of $KMnO_4$ solution with $Na_2C_2O_4$, the acidity and temperature of the solution should be strictly controlled. Does acidity too high or too low affect the determination of $KMnO_4$ concentration? Does temperature too high or too low affect the determination of $KMnO_4$ concentration?

3. When the concentration of $KMnO_4$ solution is standardized, the pink fades slowly after the first drop of $KMnO_4$ is added, and then the color fades faster. Why?

2.6 Determination of Chloride in Water by Mohr Method

AIMS

1. To learn the principle of precipitation titration.

2. To master the preparation and calibration of AgNO₃ standard solution.

3. To know the use of indicators in Mohr method.

INTRODUCTION

Mohr method is one of the significant methods for determination of chloride in water. It is also known as Argentometric method. Chloride ion is a negatively charged ion. This method is appropriate for neutral or slightly alkaline water. Water sample is titrated against standard AgNO₃ solution by using potassium chromate indicator. It is a precipitation titration method.

Generally, chloride ions are present in water as a different form of salts. The most common salts are $NaCl$, KCl, $MgCl_2$ and $CaCl_2$. They are extremely soluble in water. The sources of chloride in water may be natural or human beings. The natural sources are surrounding rock or soil or seawater intrusion in coastal areas. Whereas, various human sources are fertilizers, road salting, wastewater from industries, animal feeds, septic tank effluents etc.

The Mohr method for determination of chloride in water is a pH sophisticated method. It must be performed between the pH levels 6.5—9.0. It is better to carry out between the pH ranges 7—8. However, if ammonium buffer solution is used, the pH range of the solution is between 6.5 and 7.2. At upper pH level, the silver ions react with hydroxide ions and precipitated as silver hydroxide. In contrast, at lower pH level, potassium chromate may be converted into potassium dichromate ($K_2Cr_2O_7$) and mask the endpoint. Consequently, accurate results cannot be obtained. If the water sample is acidic, then gravimetric method or volhard's method is appropriate.

$$Ag^+ (aq) + OH^- (aq) \Longrightarrow Ag(OH)(s) \downarrow$$

$$2CrO_4^{2-} (aq) + 2H^+ \Longrightarrow Cr_2O_7^{2-} (aq) + H_2O$$

The amount of chloride in water can be simply determined by titrating the collected water sample with silver nitrate solution by using a potassium chromate indicator. The reaction is quantitative. Because the solubility of silver chloride precipitation is lower than that of silver chromate, silver chloride precipitation is first precipitated in the solution. The $AgNO_3$ reacts with chloride ion in a 1 : 1 ratio. The result is expressed as ppm.

When silver nitrate solution is gradually added into the flask, then silver ions react with chloride ions and form silver chloride. It is precipitated in bottom of the flask. The precipitation is white in color.

$$Ag^+(aq) + Cl^-(aq) =\!=\!= AgCl(s) \downarrow \quad K_{sp} = 1.8 \times 10^{-10}$$

The endpoint of the titration takes place when all the chloride ions reacts and precipitated. Then slightly extra silver ions react with the chromate ions and form a brownish-red precipitate of silver chromate. The solubility product of silver chromate exceeded in the presence of additional silver ions, and then the precipitation occurs.

$$2Ag^+ + CrO_4^{2-}(aq) =\!=\!= Ag_2CrO_4(s) \downarrow \quad K_{sp} = 2.0 \times 10^{-12}$$

Figure 2 – 11   The color change for Mohr method.

Anions that can form insoluble compounds or complexes with Ag^+, for example, PO_4^{3-}, AsO_4^{3-}, AsO_3^{3-}, S^{2-}, SO_3^{2-}, CO_3^{2-}, $C_2O_4^{2-}$, all interfere with the determination. H_2S can be removed by boiling, SO_3^{2-} can be removed by oxidation to SO_4^{2-} method. A large number of colored ions, such as Cu^{2+}, Ni^{2+}, Co^{2+}, can affect the endpoint observation. Cations which can form insoluble compounds with indicator K_2CrO_4 also interfere with the determination, such as Ba^{2+}, Pb^{2+}. The interference of Ba^{2+} can be eliminated by adding excessive Na_2SO_4. High valence metal ions such as Al^{3+}, Fe^{3+}, Bi^{3+}, Sn^{4+} are easy to hydrolyze and precipitate in neutral or weak alkaline solution, which will interfere with the determination.

Although there are many disturbances, the operation of Mohr method is simple, and the Mohr method is still mostly used in the determination of chloride ions in general water samples.

REAGENTS AND APPARATUS

• Analytical balance, 100 mL volumetric flasks, 250 mL brown volumetric flasks, 50 mL brown burettes, 250mL conical beakers, 25 mL beakers, 50 mL volumetric pipettes, 10 mL graduated cylinder, rubber bulbs, droppers, glass rods

- Water sample, sodium chloride (NaCl), silvernitrate ($AgNO_3$), 5% potassium dichromate ($K_2Cr_2O_7$) solution, distilled and deionized water

PROCEDURES

1. Preparation and standardization of 0.1 mol/L $AgNO_3$ standard solution

About 4.2 g $AgNO_3$ is weighed in a 25 mL beaker and dissolved in a proper amount of distilled water without Cl^-. Transfer the solution into a 250 mL brown volumetric flask, and wash the beaker three times with distilled water, and each wash need to be transferred to the volumetric flask. Dilute to the mark with distilled water, shake well and store in dark.

The reference NaCl of 0.5—0.65 g is accurately weighed in a small beaker (the weighing accuracy should be to ±0.0001 g), dissolved in distilled water (excluding Cl^-), and then quantitatively transferred into a 100 mL volumetric flask. The beaker is rinsed with water several times and all the wash is transferred into the volumetric flask. Then dilute the solution to the mark and shake evenly.

Accurately remove three parts of 25.00 mL NaCl standard solution in 250 mL conical bottle, add distilled water (excluding Cl^-) 25 mL, and add 5% K_2CrO_4 solution 1 mL.

Wash a 50.00 mLbrown burette thoroughly with water, and then rinse it with distilled water and a few mLs of the prepared $AgNO_3$ solution sequently. Drain through the stopcock and then fill the burette with the $AgNO_3$ solution.

Titrate NaCl standard solution with $AgNO_3$ solution under constant swirling. The solution turns from yellow to white turbidity, and then to light red turbidity (brick red), the latter is the endpoint. Read the burette to 0.01 mL and calculate the volume used in the titration by subtracting the initial volume from the final volume of $AgNO_3$. According to the concentration of NaCl standard solution and the volume of $AgNO_3$ solution, the concentration and relative standard deviation of $AgNO_3$ solution are calculated. Refill the burette, read it, and titrate the second solution, then the third.

2. Sample analysis

Transfer 25.00 mL of water sample into a 250 mL conical beaker, add distilled water (excluding Cl^-) 25 mL and 5% K_2CrO_4 solution 1 mL, under constant swirling, titrate it with $AgNO_3$ solution until the solution changes from yellow to light red turbidity. Cl^- concentration and relative mean deviation are calculated. Refill the burette, read it, and titrate the second solution, then the third.

3. Blank measurement

25.00mL distilled water is tested by the same operation as mensioned above. Repeat three times.

The final Cl^- concentration should be deducted from the blank measurement.

Silver-containing solution should be recovered and not poured into the tank at will.

DATA TREATMENT

1. Standardization of AgNO₃ standard solution

		1	2	3
NaCl weighting	$m(NaCl)/g$			
AgNO₃ titrant	before titration/mL			
	after titration/mL			
	net volume/mL			
$c(AgNO_3)/mol \cdot L^{-1}$				
average $\bar{c}(AgNO_3)/mol \cdot L^{-1}$				
the average relative deviation				

2. Sample analysis

		1	2	3
Water sample	V/mL	25.00	25.00	25.00
AgNO₃ titrant	before titration/mL			
	after titration/mL			
	net volume/mL			
$c(Cl^-)/mol \cdot L^{-1}$				
average $\bar{c}(Cl^-)/mol \cdot L^{-1}$				
the average relative deviation				
average $\bar{c}(Cl^-)$ deduct the blank/mol $\cdot$ L^{-1}				

3. Blank measurement

		1	2	3
Distilled water	V/mL	25.00	25.00	25.00
AgNO₃ titrant	before titration/mL			
	after titration/mL			
	net volume/mL			
Blank $c(\text{Cl}^-)$/mol · L^{-1}				
Average blank $\bar{c}(\text{Cl}^-)$/mol · L^{-1}				
the average relative deviation				

QUESTIONS

1. Why should we control the pH of NaCl solution in the range between 6.5 and 10.5 when Cl$^-$ is measured by Mohr method?

2. When K_2CrO_4 is used as an indicator, will the concentration of K_2CrO_4 affect the endpoint? If yes, how?

3. Can AgNO$_3$ be put in the conical beaker, and directly titrated with NaCl standard solution by Mohr method? Why?

4. What is the significance of blank measurement?

2.7 Iodometric Titration of Vitamin C (Ascorbic Acid)

> **AIMS**
>
> 1. To learn the operation steps and precautions of direct iodometry.
> 2. To Master the principle and conditions for the determination of vitamin C.

INTRODUCTION

Vitamin C or ascorbic acid is essential for human life and is required for a range of physiological functions in human body. It can be found either in fresh fruits and vegetables naturally or in medical forms such as normal tablets, effervescent tablets and liquid vials. It is the most widely taken supplement. Though daily requirements of vitamin C are changeable according to the age, sex and conditions, it is around 75 to 90 mg per day for healthy adults and no more than 2 000 mg per day is recommended.

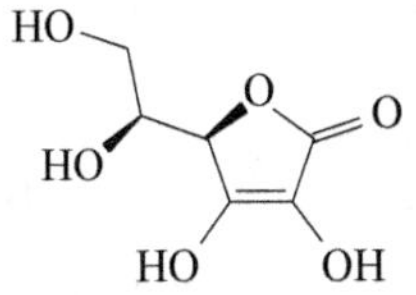

Figure 2 – 12 Chemical structure of Vitamin C

Ascorbic acid is a water soluble vitamin with molecular weight of 176. 12 g • mol^{-1} and melting point of 193 ℃. It is one of the most ubiquitous vitamins ever discovered. Ascorbic acid is a reducing agent which reverses the oxidation in aqueous solution. Increased amounts of free radicals trigger the condition called oxidative stress which is kept under control by antioxidants. If there are not enough antioxidants some stress related diseases including hypertension, atherosclerosis, chronic inflammatory diseases and diabetes might occur. Besides playing a paramount role as an antioxidant and free radical scavenger, it has been suggested to be an effective antiviral agent. In addition, ascorbic acid has been widely used in the pharmaceutical, chemical, cosmetic and food industry as antioxidant. Therefore, there is a need to find an accurate, reliable, rapid, and easy-to implement method for measuring the amount of ascorbic acid in a sample.

However, there have been difficulties in quantifying ascorbic acid due to its instability in aqueous solution. The instability of ascorbic acid is due to its oxidation to dehydroascorbic acid, which is a reversible reaction, and subsequently to 2, 3-diketo-L-gulonic acid. The later reaction is irreversible. Numerous analytical techniques are proposed for the determination of vitamin C in different matrices. Some of the techniques include: direct titration, fluorometric methods, chromatographic methods, and Electrochemical 7, 14—18. However, some of these methods are time-consuming, some are costly, some need special training operators, or they suffer from the insufficient sensitivity or selectivity.

Titration method to determine vitamin C is used in this experiment because it is a quick, accurate and precise method. Triiodide, I_3^-, is a mild oxidizing agent that is widely used in oxidation/reduction titrations. Triiodide is prepared by combining potassium iodide, KI, and potassium iodate, KIO_3, in acidic solution according to the following stoichiometry:

$$IO_3^- + 8I^- + 6H^+ \Longrightarrow 3I^- + 3H_2O \qquad (2\text{-}7\text{-}1)$$

In preparing triiodide, excess KI is used, so the concentration of I_3^- is determined by the amount of KIO_3 added to the solution. Triiodide reacts with ascorbic acid (vitamin C, a mild reducing agent) to form dehydroascorbate and three iodide ions according to the reaction:

$$H_2O + I_3 - + \text{[structure]} = \text{[structure]} + 3I^- + 2H^+ \qquad (2\text{-}7\text{-}2)$$

Because of its strong reducibility, Vc is easily oxidized by oxygen in solution and air, which is stronger in alkaline medium, so titration should be carried out in acidic medium to reduce the occurrence of side reactions. Notice that one mole of iodine is consumed for each mole of ascorbic acid. In this experiment, you will determine the amount of ascorbic acid in a vitamin pill using the triiodide reaction in a "back titration". After extracting the ascorbic acid from vitamin pills with acid, you will convert it to dehydroascorbate using a known excess of triiodide. The amount of triiodide remaining after reaction (2-7-2) will be determined by titration of the triiodide with a standardized thiosulfate solution. Note that you do not titrate the analyte directly, but rather titrate an added reagent after excess has been added. This is known as a back titration. The back titration reaction is:

$$I_3^- + 2S_2O_3^{2-} = 3I^- + S_4O_6^{2-} \qquad (2\text{-}7\text{-}3)$$

Note that 2 moles of thiosulfate are consumed for each mole of triiodide present. The endpoint is determined using a starch indicator. Mixtures of starch and triiodide have a deep violet color, but when the triiodide is consumed the solution becomes colorless. Over time the starch-triiodide complex can stabilize, and it becomes difficult to reduce all of the triiodide. Therefore it is preferable to add the starch just before the endpoint. Fortunately the triiodide solution itself has a yellow-to-brown color, depending on concentration. When the solution turns pale yellow, you know that most of the triiodide has been consumed, and you are near the endpoint. Then you can add the starch indicator. You know how much I_3^- is added to the vitamin sample, and with the titration results you can determine how much is left after the oxidation of ascorbate. The difference between these is the amount of triiodide consumed in the oxidation of ascorbate, which is related to the amount of vitamin C present in the sample by the stoichiometry of reaction (2-7-2).

REAGENTS AND APPARATUS

- Analytical balance, 250 mL volumetric flasks, 50 mL burettes, 250 mL conical beakers, 250 mL beakers, 25 mL volumetric pipettes, 10 mL, 25 mL, and 100 mL graduated cylinders, droppers, glass rods, rubber bulbs, mortars and pestles

- Vitamin tablets, potassium iodate (KIO_3), potassium iodide (KI), sodium thiosulfate ($Na_2S_2O_3$), 0. 3 mol $\cdot$ L^{-1} sulfuric acid (H_2SO_4), 1% starch indicator solution, distilled and deionized water.

PROCEDURES

1. Preparation of 0.01mol $\cdot$ L^{-1} KIO_3 solution

Accurately weigh approximately 0. 54 g of solid reagent and record the mass to 4 decimal places. Deliver the KIO_3 to a 250 mL volumetric flask and add 100 mL of distilled water. Swirl to dissolve, then dilute to volume. Compute the molarity of the solution. (FW=214. 00)

2. Standardizing the thiosulfate solution

Weigh approximately 0. 4 g $Na_2S_2O_3$ and record the mass to 4 decimal places. Dissolve the thiosulfate in a 250 mL beaker with 100 mL distilled water, and then quantitatively transferred into a 250 mL volumetric flask. Dilute the solution to the mark and shake it evenly. You must use the same solution for the entire experiment.

Pipet 25. 00 mL of the KIO_3 solution into each of 3 conical beakers. Using the calculated KIO_3 solution concentration, calculate the volume of titrant (thiosulfate) required assuming that the thiosulfate concentration is 0. 01 mol $\cdot$ L^{-1}. This gives the approximate endpoint. Add 1 g of KI and 20 mL of 0. 3 mol $\cdot$ L^{-1} sulfuric acid solution to each conical beaker.

Titrate the triiodide with the thiosulfate solution until the brown solution becomes pale yellow. Then add 2 mL of the starch indicator solution and titrate until the violet color of the starch-iodine complex just disappears. This is the endpoint. Record the initial and final reading to 2 decimal places. Repeat this procedure for a total of three precise titrations.

3. Analyzing the vitamin C

Weigh a sufficient number of vitamin tablets so that approximately 500 mg of ascorbicacid is obtained (normally one tablet contains 50—250 mg vitamin C, and your TA may tell you how many tablets to use). Grind the tablets with a mortar and pestle.

Record the weight to 4 decimal places of the resulting powder that is actually analyzed. Dissolve the known mass of powder in a 250 mL beaker with 100 mL 0. 3 mol $\cdot$ L^{-1} sulfuric acid. Then transfer to a 250 mL volumetric flask. Rinse the beaker and three times with 0. 3 mol $\cdot$ L^{-1} sulfuric acid, and all the rinse need to be tranferred to the 250 mL volumetric flask. Swirl the flask for about 10 minutes, and then let it stand for several minutes. Swirl again, and then dilute to the mark with 0.3 mol·L⁻¹ sulfuric acid. Because of the fillers and binders used in vitamin tablets, your solution may be cloudy, and this is normal.

Deliver 25.00 mL of the vitamin C solution to an conical beaker. Add 1 g of solid KI and 25.00 mL of standardized KIO₃ to the flask. Titrate the remaining triiodide with the standardized thiosulfate solution as above, taking care to add the starch solution just before the endpoint. Record the initial and final reading to 2 decimal places. Repeat this titration twice for a total of three precise determinations.

Calculate the average mass of vitamin C in each tablet and the uncertainty in the determination.

DATA TREATMENT

1. Preparation of 0.01mol·L⁻¹ KIO₃ solution

KIO₃	Mass/g	Volume/mL	Molarity/mol·L⁻¹
		250.0	

2. Standardizing the thiosulfate solution

		1	2	3
$Na_2S_2O_3$ weighting	$m(Na_2S_2O_3)/g$			
$Na_2S_2O_3$ titrant	before titration/mL			
	after titration/mL			
	net volume/mL			
$c(Na_2S_2O_3)/mol·L^{-1}$				
average $\bar{c}(Na_2S_2O_3)/mol·L^{-1}$				
the average relative deviation				

4. Analyzing the vitamin C

		1	2	3
Vc weighting	$m(Vc)/g$			
$Na_2S_2O_3$ titrant	before titration/mL			
	after titration/mL			
	net volume/mL			
$c(Vc)/mol·L^{-1}$				
average $\bar{c}(Vc)/mol·L^{-1}$				
the average relative deviation				

QUESTIONS

1. Adding some KI when preparing the iodine solution can speed up the dissolution of I_2. Why? Does KI have any effect on the concentration of the iodine solution?

2. Why should dilute sulfuric acid be added to the Vc solution?

3. What are the sources of error in the iodometry? What measures should be taken to reduce errors?

2.8 Quantitative Determination of Sulfate by Gravimetric Analysis

AIMS

1. To know the principle of gravimetric analysis.

2. To master the operation procedure of gravimetric analysis.

INTRODUCTION

Gravimetric analysis is based on the measurement of the mass of a substance of known composition that is chemically related to the analyte. Gravimetric analysis includes precipitation, volatilization and electrodeposition methods.

In precipitation gravimetry of the analyte is carried out by the use of inorganic or organic precipitating agents. The two common inorganic precipitating agents are silver nitrate, which is used to precipitate halide ions such as chloride, and barium chloride for precipitating sulfate ions. Additionally, potassium, ammonium, rubidium, and cesium ions can be precipitated by sodium tetraphenylborate.

Sulfate is quite common in nature and may be present in natural water in concentrations ranging from a few to several thousand milligrams/liter. Sulfates are of considerable concern because they are indirectly responsible for two serious problems associated with the handling and treatment of wastewater. Odor and sewer corrosion problems result from the reduction of sulfates to hydrogen sulfide under anaerobic conditions.

In an aqueous solution, the sulfate ion undergoes the following reaction with barium:

$$Ba^{2+}(aq)+SO_4^{2-}(aq)\Longrightarrow BaSO_4(s)$$

$$K_{sp}=[Ba^{2+}] \cdot [SO_4^{2-}]=1.1\times10^{-10} \quad \text{at } 25\ ℃$$

Solubility of $BaSO_4$ at room temperature is around 0.3—0.4 mg per 100 g of water. Its solubility increases when excessive amount of mineral acid is present. On the other hand, precipitation should be done in acidic medium. Because in neutral and basic solutions Ba^{2+} ions precipitate with PO_4^{3-}, CO_3^{2-} or OH^- ions which are present in the solution. Therefore, precipitation is carried out in weakly acidic medium and addition of excess acid is avoided. Precipitation in a weakly acidic medium provides precipitate to occur in the form of large crystalline particles.

Barium sulfate is collected on a suitable filter, washed with water, then ignited and weighed. From the mass of $BaSO_4$, the amount of sulfate present in the original sample is calculated. Barium sulfate usually precipitates as very fine particles. The high surface area of the particles facilitates contamination by adsorption. Larger particles can be obtained by heating the precipitate in the presence of its mother liquor. During this digestion process, recrystallization takes place resulting in a precipitate of larger particle size. Since barium sulfate is stable in air and is nonhygroscopic, the weighing can be performed in an open crucible.

Recrystallized $BaSO_4$ is collected on filter paper. The precipitate and filter paper are dried in a crucible. The filter paper is charred and ashed, leaving the dried sample in the crucible for ease in weighing. To reduce errors, the paper ashing operation must be carried out with care. The paper should be charred carefully at low temperature with the crucible lid in place so that escaping gases do not burst into flame. When smoke formation has ceased, the temperature is slowly increased and air is allowed ample access to the interior of the crucible. If this practice is not followed, carbon will reduce sulfate ion to sulfide ion, forming carbon monoxide:

$$BaSO_4+4C\Longrightarrow BaS+4CO\uparrow$$

Low mass determinations will result.

From the mass of $BaSO_4$, the amount of sulfate present in the original sample is calculated.

Although this method appears to be rather straightforward, variations in the acidity, temperature, manner of addition of the precipitant and time of digestion markedly affect the filterability of the barium sulfate precipitate and the extent to which various foreign ions are coprecipitated. Foreign anions such as nitrate, chlorate and chloride are coprecipitated as their barium salts, and the ignited precipitate contains the salt or oxide as an additive impurity. The coprecipitation of chloride can be decreased by slow addition of the precipitant. Since nitrate and chlorate interfere even at low concentrations, they should be removed from the solution before precipitation.

Foreign cations such as ferric iron, calcium, and to a lesser extent, the alkali metals are coprecipitated as the sulfates. These are substitutional impurities, and the magnitude of the error depends upon the differences between the weight of the foreign sulfate or oxide and the weight of an equivalent amount of barium sulfate. The presence of ferric iron can produce errors as high as 2% in the determination.

This precipitation gravimetry can be used to determine sulfate in surface water, groundwater, saline water, domestic sewage and industrial wastewater. The color of water sample does not affect the determination. Water samples with sulfate content above $10 \ mg \cdot L^{-1}$ can be determined, and the upper limit of determination is $5\ 000\ mg \cdot L^{-1}$.

Table 2 − 3   The common interferences affecting the sulfate analysis.

Effect on analysis	Nature of interference
Low Results	• Excess amounts of mineral acid present. • Coprecipitation of sulfuric acid. Note that this is a source of error in a gravimetricdetermination of sulfate but not of barium, since this $H_2 SO_4$ is driven off during ignition. • Coprecipitation of alkali metal and various divalent ions. Sulfates of these ions usually weigh less than the equivalent amount of $BaSO_4$, which have formed. • Coprecipitation of ammonium ion, $(NH_4)_2 SO_4$, which is volatilized upon ignition of the precipitate. • Coprecipitation of iron as a basic iron (III) sulfate. • Partial reduction of $BaSO_4$ to BaS when filter paper charred too rapidly. • In the presence oftrivalent chromium, complete precipitation of $BaSO_4$ may not be achieved owing to formation of soluble sulfates of chromium (III).
High Results	• Absence of mineral acid. Slightly soluble carbonate or phosphate of barium may precipitate. • Coprecipitation of barium chloride. • Coprecipitation of anions, particularly nitrate and chlorate, in the form of barium salts.

REAGENTS AND APPARATUS

- Analytical balance, water bath, water pump, muffle furnace, bunsen burner, crucibles with lids, desiccator, ringstand, wire gauze, tong, 400 mL beakers, 100 mL graduated cylinder, brinell funnel, 500 mL filter bottle, ashless filter paper, glass rods, watch glass, adapters
- $6.0 \ mol \cdot L^{-1}$ concentrated hydrochloric acid (HCl), $0.10 \ mol \cdot L^{-1}$ barium chloride ($BaCl_2$) solution, unknown sulfate solution, $0.10 \ mol \cdot L^{-1}$ iron nitrate ($Fe(NO_3)_3$) solution

PROCEDURES

1. Preparation of Crucibles

Each crucible should be cleaned and rinsed thoroughly with distilled water. Make sure that the crucibles are marked properly so they can be distinguished from one another. Use a permanent marker, not a paper or tape label. You can mark the sides of crucibles with a solution of iron nitrate.

For drying, place the cleaned crucibles in the furnace. Remove the crucibles with tongs (never touch crucibles with your hands or with paper for the duration of the experiment) and allow them to cool for 5 minutes before placing them in a desiccator for cooling to room temperature. Cooling will take about 10 min in the desiccator. Weigh crucibles to the nearest 0.000 1 g. Return them to the oven for 1 hour and repeat the weighing process which should be carried out until two consecutive masses agree to within 0.0010 g. It is extremely important that the crucibles should be treated exactly in the same way during this process and later on when they contain the precipitate.

Note: You need to use the same balance throughout the course of this experiment. Use of different balances, when weighing the crucibles, will introduce an error into your calculations (a common cause for not being able to bring the crucibles to constant mass).

2. Preparation and Precipitation of the Unknown Samples

Obtain three replicate unknown sample solutions. Add 100 mL of distilled water using a graduated cylinder to each solution in the beakers. Add 4.0 mL of 6.0 mol $\cdot$ L^{-1} HCl, cover the beaker with watch glass and heat the solution nearly to boiling in a water bath.

For each sample, heat 50.0 mL of 0.10 mol $\cdot$ L^{-1} BaCl$_2$ solution in a beaker nearly to boiling. Add this solution quickly with vigorous stirring to the hot sample solution. Use a separate stirring rod for each sample and leave it in the solution throughout the experiment. Rinse the beaker walls with distilled water and then cover them with a watch glass. Digest the precipitated BaSO$_4$ at just below the boiling point for 2 hours in the water bath.

Decant the hot supernatant through a fine ashless filter paper placed on a Brinell funnel. Make sure the filter paper is well-seated. Initially, filter as much of the supernatant liquid as possible (solid accumulating on the filter paper drastically slows the rate of filtration). Rinse the glass rod and the beaker with distilled water to recover the final pieces of precipitate. Wash the precipitate twice by using about

10. 0 mL portions of distilled water for each wash.

Place paper and its contents into a porcelain crucible that has been brought to a constant mass previously. Gently char off the paper on a Bunsen burner: Place the crucible vertically on a triangle supported by a ring stand and adjust the ring so that the bottom of the crucible is positioned 10 to 15 cm above a flame which is 1 to 2 cm in height as shown in Figure 2 – 13.

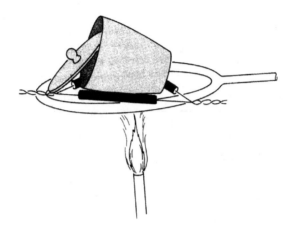

Figure 2 – 13   Igniting a precipitate.

Place the lid on the crucible but displace it to one side so that steam can escape through a slit of ∼2 mm in width. Apply heat slowly and gently so that violent boiling of the water and bursting of the package avoided. When drying is complete, fully cover the crucible and char the paper by increasing the heat applied to the crucible. Escaping gases should not burst into the flame.

Occasionally lift the lid and check the progress of the charring operation, by observing the blackening of the paper and the disappearance of white areas. Because of difficulty of drying and weighing a precipitate on a filter paper, it is burned away, leaving behind only the precipitate.

• At high temperatures, $BaSO_4$ may be reduced to BaS by the reaction with C of the filter paper, which can be prevented by burning the filter paper at rather low temperatures.

Ignite the crucible to a constant mass at 800 ℃ in an electric furnace, for 1 hour. Cool and weigh. Repeat heating, cooling and weighing until the mass of the crucible is constant within ± 0. 001 0 g. Once a constant mass is reached, discard the solid in the waste container provided.

The term ignition means "to heat to a high temperature" not "to set to fire to" Ifignition is done at very high temperature BaSO4 may decompose as follows.

$$BaSO_4(s) = BaO(s) + SO_3(g) \uparrow$$

Clean the crucibles by rinsing each thoroughly with distilled water and return them to the technician.

DATA TREATMENT

1. Preparation of Crucibles

Crucible label	1	2	3
Weight (initial cooling)/mg			
Weight(consecutive cooling)/mg			

2. Preparation and Precipitation of the Unknown Samples

Crucible label	1	2	3
Weight (after filter)/mg			
Weight (initial cooling)/mg			
Weight (consecutive cooling)/mg			
Sulfate in each unknown sample/mg			
Meansulfate value/mg			
% relative standard deviation			

QUESTIONS

1. If ordinary filter paper, instead of ashless paper were used, how would your experimental results be affected? Would they be too high or too low?

2. What is the importance of digestion step during precipitation?

3. What is the importance of making the sulfate solution slightly acidic before the addition of $BaCl_2 \cdot 2H_2O$ solution?

4. What is the importance of ignition at proper temperature?

5. What are the mostimportant errors in this procedure?

2.9 Determination of Dissociation Degree and Dissociation Constant of Acetic Acid

AIMS

1. To learn the basic principles and methods of determining dissociation degree and dissociation constant of acetic acid.

2. To know how to use a pH meter.

3. To review the use of volumetric flask and pipettes.

INTRODUCTION

Acids and bases play a significant role in many areas of chemistry and biochemistry. We can classify substances as acids and bases based on chemical behavior. The definition of an Arrhenius acid is a substance the produces hydronium ions (H_3O^+) in aqueous solution, while an Arrhenius base is a substance that produces hydroxide ions (OH^-). A more useful definition of acids and bases is based on the Brønsted-Lowry theory. A Brønsted-Lowry acid is defined as a substance that acts as a proton donor, while a Brønsted-Lowry base acts as a proton acceptor.

Consider the behavior of hydrochloric acid illustrated in Eq. (2-9-1)

$$HCl(aq) + H_2O(l) \longrightarrow H_3O^+(aq) + Cl^- \qquad (2\text{-}9\text{-}1)$$

In this reaction, the acid HCl donates a proton (i. e. , H^+) to a water molecule (the base) to produce the ions H_3O^+ and Cl^-. HCl is a strong acid, which means that ionization is essentially complete; nearly 100% of the HCl in solution has been converted to H_3O^+ and Cl^- ions.

By contrast, weak acids are substances that only ionize to a slight extent. Consider the ionization of acetic acid (CH_3COOH), the major component of vinegar, shown in Eq. (2-9-2)

$$CH_3COOH(aq) + H_2O(l) \leftrightarrow H_3O^+(aq) + CH_3COO^-(aq) \qquad (2\text{-}9\text{-}2)$$

In this case only about 1% of the acetic acid dissociates into ions. An equilibrium state, indicated by the double-headed arrow, exists between the undissociated acid on the left and the ions on the right. The extent of dissociation is indicated by the value of the acid dissociation constant, or K_a, which is calculated as:

$$K_a = \frac{[H_2O^+][CH_3COO^-]}{[CH_3COOH]} \qquad (2\text{-}9\text{-}3)$$

The square brackets ([ ]) indicate the molar concentrations of the products and reactants are at equilibrium. Although water appears as a reactant, its concentration remains essentially constant at 55.5 M in dilute solutions and is not included in the expression for K_a. If we know the initial concentration of acetic acid and the extent of ionization, then we can calculate the equilibrium concentrations of the acetic acid and acetate radical, and use Eq. (2-9-3) to calculate K_a. This approach is demonstrated in Example 1.

Example 1. A 0.12 M solution of unknown weak acid has a pH=3.38. Calculate the value of K_a for this acid. We can derive expressions for the equilibrium concentration by setting up a reaction table as follows:

$$HA(aq) + H_2O(l) \leftrightarrow H_3O^+(aq) + A^-(aq)$$

Initial conc. :   0.12 M   —   —

Change: $-x$ $+x$ $+x$

At equilibrium: $(0.12-x)$ (x) (x)

Substituting these equilibrium expressions into the expression for K_a yields

$$K_a = \frac{[x][x]}{[0.12-x]} \qquad (2\text{-}9\text{-}4)$$

If we know the value of x, we can calculate K_a. Since the pH of the solution is known, we can use the pH to calculate $[H_3O^+]$ using Eq. (2-9-5). Since $[H_3O^+]$ is equal to x in our reaction table, we can now calculate K_a.

$$[H_3O^+] = 10^{-pH} = 10^{-3.38} = 4.17 \times 10^{-4} \qquad (2\text{-}9\text{-}5)$$

And

$$K_a = \frac{[x][x]}{[0.12-x]} = \frac{[4.17 \times 10^{-4}]^2}{\sim 0.12} = 1.45 \times 10^{-6} \qquad (2\text{-}9\text{-}6)$$

Note that the value of x in Example 1 is negligible compared to the initial concentration of the weak acid, so it can be ignored in the denominator of the K_a expression without introducing significant error in the result. If the value of K_a is $> 10^{-3}$ the extent of dissociation may not be negligible, however, and a more rigorous mathematical treatment may be necessary.

The dissociation degree is the fraction of original solute molecules that have dissociated. It is usually indicated by the Greek symbol α. More accurately, the degree of dissociation refers to the amount of solute dissociated into ions or radicals per mole.

$$\alpha = \frac{c(H^+)}{c} \times 100\% \tag{2-9-7}$$

A pH Meter is a scientific instrument that measures the hydrogen-ion concentration (or pH) in a solution, indicating its acidity or alkalinity. The pH meter measures the difference in electrical potential between a pH electrode and a reference electrode. It usually has a glass electrode plus a calomel reference electrode, or a combination electrode. In addition to measuring the pH of liquids, a special probe is sometimes used to measure the pH of semi-solid substances.

For very precise work the pH meter should be calibrated before each measurement. For normal use calibration should be performed at the beginning of each day. The reason for this is that the glass electrode does not give a reproducible Electromotive force over longer periods of time.

Calibration should be performed with at least two standard buffer solutions that span the range of pH values to be measured. For general purposes buffers at pH 4. 00 and pH 10. 00 are acceptable. The pH meter has one control (calibrate) to set the meter reading equal to the value of the first standard buffer and a second control which is used to adjust the meter reading to the value of the second buffer. A third control allows the temperature to be set.

For more precise measurements, a three buffer solution calibration is preferred. As pH 7 is essentially, a "zero point" calibration (akin to zeroing or taring a scale or balance), calibrating at pH 7 first, calibrating at the pH closest to the point of interest (e. g. either 4 or 10) second and checking the third point will provide a more linear accuracy to what is essentially a non-linear problem. The calibration process correlates the voltage produced by the probe (approximately 0. 06 volts per pH unit) with the pH scale. After each single measurement, the probe is rinsed with distilled water or deionized water to remove any traces of the solution being measured, blotted with a scientific wipe to absorb any remaining water which could dilute the sample and thus alter the reading, and then quickly immersed in a solution suitable for storage of the particular probe type.

REAGENTS AND APPARATUS

- pH meter, 20. 00 mL measuring pipettes, 50 mL volumetric flasks, 50 mL beakers, rubber bulbs, glass rods.
- 0. 1 mol $\cdot$ L^{-1} acetic acid (HAc) solution, phosphate standard buffer solution (pH 6. 86 at 25 ℃), potassium hydrogen phthalate (KHP) standard buffer solution (pH 4. 01 at 25 ℃), distilled water

PROCEDURES

1. Preparation of acetic acid solutions with different concentrations

Transfer 0.1 mol $\cdot$ L^{-1} HAc solution of 3.00 mL, 6.00 mL and 15.00 mL into three 50 mL volumetric flasks by measuring pipette, respectively. Dilute the solution to the mark with distilled water, cover the glass stopper, and invert it up and down ten times to shake well. Four solutions with different concentrations can be obtained, three prepared and one undiluted HAC solution. Label them 1, 2, 3 and 4 from the dilute to the concentrated.

2. Calibrate the pH meter

Turn on the pH meter and allow it to warm up for several minutes. Carefully rinse the glass electrode in the pH meter with distilled water from a wash bottle. Then obtain 20 mL of the phosphate buffer solution in a clean, dry 50 mL beaker. Measure the temperature of the buffer solution, and set the temperature compensation knob on the pH electrode to match the solution temperature. Carefully immerse the pH electrode in the phosphate buffer solution and turn the function knob to "pH". Allow the pH electrode reading to stabilize, then adjust the pH display until the pH reading matches the pH of your standard solution.

Obtain 20 mL of the KHP buffer solution in a clean, dry 50 mL beaker. Rinse the glass electrode again. Carefully immerse the pH electrode in the KHP buffer solution and turn the function knob to "Slope". Adjust the pH display until the pH reading matches the pH of the KHP buffer solution, the calibration is complete.

Turn the function knob to "Standby". Carefully remove the pH electrode from the buffer solution and rinse it with distilled water from a rinse bottle.

3. Determination of pH of acetic acid solution

Pour about 20 mL of Solution 1 into a clean, dry 50 mL beaker. Place the pH electrode in this solution and turn the function knob to "pH". Read the pH of the solution to the nearest 0.01 pH unit and record this pH value on your Data Sheet. Rinse the pH electrode with distilled water. Repeat these steps for Solution 2—4.

Discard the solution in the beaker and in the volumetric flask as instructed. Wash and rinse the beaker and volumetric flask with distilled water.

DATA TREATMENT

	$V(HAc)/$ mL	$V(H_2O)/$ mL	$c(HAc)/$ mol $\cdot$ L^{-1}	pH	$c(H^+)/$ mol $\cdot$ L^{-1}	α %	K_a
1							
2							
3							
4							
5							
The average value of K_a							

QUESTIONS

1. Do the beakers used in this experiment have to be dried?

2. If a part of the solution is taken out by the glass rod after mixing, does it affect the results?

3. When measuring the pH value of HAc solution, why should we take the order of concentration from the dilute to the concentrated?

2. 10 Determination of Catalyzed Decomposition Rate Constants of Hydrogen Peroxide

AIMS

1. To learn the characteristics of first-order reactions.

2. Determination of rate constants and series of decomposition of hydrogen peroxide.

3. To learn the effects of various factors on the reaction rate.

4. How to obtain the catalyzed decomposition rate constants of hydrogen peroxide from graphic method.

INTRODUCTION

Any molecule in motion possesses kinetic energy, and the faster it moves, the greater the kinetic energy it has. When molecules collide, some of the energy is converted into vibrational energy. If the vibrational energy is large, it may cause some of the chemical bonds in the molecule to break. Breaking bonds is the first step towards product formation. If the initial kinetic energies are small, the molecules will merely bounce off each other without breaking any bonds. In order to react, the molecules must have a total kinetic energy equal to, or greater than, the activation energy, E_a. The activation energy is the minimum amount of energy required to initiate a chemical reaction.

The energy change in a reaction is given on an energy level diagram, or potential energy profile, as shown in Figure 1. The vertical axis gives the potential energy for the reaction, while the horizontal axis is arelative (i. e. , time) scale that shows the progress of the reaction. The diagram indicates that there is a "hill" or energy barrier that needs to be overcome before any products can be formed. If the collision between the molecules produces enough energy to overcome the barrier, the reactant molecules are in a temporary transition state, forming an activated complex at the height of the barrier before forming the product molecules.

The change in concentration and temperature affect the rate of a chemical reaction by influencing the collisions among the molecules. An increase in concentration increases the number of molecular collisions. An increase in temperature increases the rate because the molecules move faster, increasing the kinetic energy and collisions are more frequent.

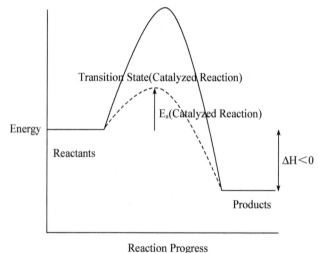

Figure 2 - 14   Progress of reaction

A catalyst is a compound that affects the rate of chemical reactions by lowering the activation energy, E_a. With a lower activation energy, less energy is needed to overcome the energy barrier to form the product (s). A catalyst takes part in the reaction, but is not consumed. It always returns to its original composition at the end of the reaction. Catalysts are defined as homogeneous or heterogeneous where the catalyst is either in the same phase as the reactants or in a different phase, respectively.

Consider a chemical system in which there are two reactants, A and B, with stoichiometric coefficients a and b, forming c moles of product C.

$$aA+bB \longrightarrow cC \qquad (2\text{-}10\text{-}1)$$

The rate of the reaction can bedefined as the change in concentration of the product as a function of the change in time:

$$\text{Rate}=\frac{\Delta c}{\Delta t} \qquad (2\text{-}10\text{-}2)$$

Although the rate of the reaction is constant under the conditions we set, we can vary it between trials by changing the initial concentrations of the reactants. For a given reaction, the rate typically increases with an increase in the concentration of a reactant. For our general reaction, the relationship between the rate and the concentration of the reactants is given by the rate law

$$\text{Rate}=k[A]^m[B]^n \qquad (2\text{-}10\text{-}3)$$

In this rate law equation, k is a constant, called the rate constant of the reaction and should be the same for trials of the reaction at a given temperature. The superscripts, m and n, are the order of the reaction with respect to A and B and are integers, such as 0, 1, 2, or 3. The values of m and n do not come from the coefficients of the balanced chemical equation, although they may occasionally have the same value. For example, if m is 1, the reaction is said to be first order with respect to reactant A. The sum of orders (i. e. , $m+n$) is the overall order of the reaction. The order of a reaction must be determined experimentally and cannot be determined from the chemical equation alone.

Determining the order of the reaction must be done using experimental data. Pairs of trials can be compared that vary only the concentration of one reactant. If changing the concentration of that reactant causes a change in the rate, then we can say that the rate is dependent to some degree on the concentration of that species. Conversely, if the rate does not change with the change in concentration of one of the reactants, the rate is not dependent on the concentration of that species.

If the concentration of the reactant doubles and the initial rate ...	then the reaction is _____ order with respect to that reactant.
doubles(2^1)	first
quadruples(2^2)	second
increases eight-fold (2^3)	third
If the concentration of the reactant triples and the initial rate ...	then the reaction is _____ order with respect to that reactant.
triples(3^1)	first
increases nine-fold (3^2)	second

The decomposition of hydrogen peroxide in aqueous solution proceeds very slowly. A bottle of 3% hydrogen peroxide sitting on a grocery store shelfis stable for a long period of time. The decomposition takes place according to the reaction below.

$$2H_2O_2(aq) \leftrightarrow 2H_2O + O_2(g) \tag{2-19-4}$$

A number of catalysts can be used to speed up this reaction, including potassium iodide, manganese (IV) oxide, and the enzyme catalase. If you conduct the catalyzed decomposition of hydrogen peroxide in a closed vessel, you will be able to determine the reaction rate as a function of the pressure increase in the vessel that is caused by the production of oxygen gas. If you vary the initial molar concentration of the H_2O_2 solution and the catalyst (KI) concentration, the rate law for the reaction can also be determined. Finally, by conducting trials at different temperatures with the same concentrations, the activation energy, E_a, can be calculated.

REAGENTS AND APPARATUS

- Lab Pro, gas pressure sensor, thermometer, tubing with two Luer-lock connectors, one-hole rubber stopper with stem, solid rubber stopper, test tubes, 1 L beakers, 10 mL graduated cylinders, plastic Beral pipet.
- 3% hydrogen peroxide (H_2O_2) solution, 0.5 mol·L^{-1} potassium iodide (KI) solution, distilled water.

PROCEDURES

1. Decompose 3% H_2O_2 solution with 0.5 mol·L^{-1} KI solution at 20 ℃

Obtain and wear goggles. Prepare the reagents for temperature equilibration. Obtain room-temperature water to set up a water bath to completely immerse the test tube. Use a thermometer or a temperature probe to measure the temperature of the bath. Record this temperature in your data table; presume that the water bath temperature remains constant throughout.

Measure out 4 mL of 3% H_2O_2 solution into the test tube. Seal the test tube with the solid rubber stopper and place the test tube in the water bath. Measure out 2 mL of 0.5 mol $\cdot$ L^{-1} KI solution in a graduated cylinder. Draw 1 mL of the KI solution into a graduated Beral pipet. Invert the pipet and immerse the reservoir end of the pipet in the water bath.

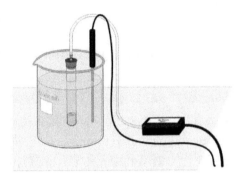

Figure 2 – 5   Reaction and measurement device

Connect the gas pressure sensor to Lab Pro and choose "New" from the file menu. If you have an older sensor that does not auto-ID, manually set up the sensor. Use the plastic tubing to connect the one-hole rubber stopper to the gas pressure sensor, as shown in Figure 2. About one-half turn of the fittings will secure the tubing tightly. On the Meter screen, tap Rate. Change the data-collection rate to 0.1 samples $\cdot$ sec^{-1} and the data-collection length to 300 seconds and select OK.

Prepare to run the reaction and collect pressure data. Remove the test tube from the water bath and remove the solid stopper. Remove the plastic Beral pipet from the water bath and quickly transfer the 1 mL of KI solution into the test tube. Tap or lightly shake the test tube to mix the reagents. Seal the test tube with the one-hole stopper connected to the gas pressure sensor. Place the test tube back in the water bath.

Start data collection. Data will be collected for five minutes. If necessary, gently hold the test tube so that it stays completely immersed in the water bath. When the data collection is complete, carefully remove the stopper from the test tube to relieve the pressure. Dispose of the contents of the test tube as directed.

Rinse and clean the test tube for the next measurement.

2. Decompose 3% H_2O_2 solution with 0.25 mol $\cdot$ L^{-1} KI solution at 20 ℃

Measure out 4 mL of 3% H_2O_2 solution into the test tube. Seal the test tube with the solid rubber stopper and place the test tube in the water bath. Add 1 mL of distilled water to the remaining 1 mL of KI solution in the graduated cylinder. Swirl the mixture gently to mix the solution.

Draw 1 mL of the KI solution into a plastic Beral pipet. Invert the pipet and immerse the reservoir end of the pipet in the water bath. Allow both the test tube and the Beral pipet to remain in the water bath for at least two minutes before proceeding.

Repeat the collect pressure data steps mentioned above in Step 1. Remember to save and record the data.

3. Decompose 1.5% H_2O_2 solution with 0.5 mol • L^{-1} KI solution at 20 ℃

Prepare a 1.5% H_2O_2 solution by mixing 2 mL of distilled water with 2 mL of 3% H_2O_2 solution. Transfer all 4 mL of the 1.5% H_2O_2 solution to the test tube, seal the test tube with the solid stopper, and place the test tube in the water bath.

Rinse and clean the graduated cylinder that you have used for the KI solution. Add a fresh 2 mL of 0.5mol • L^{-1} KI solution to the graduated cylinder. Draw 1 mL of the KI solution into a plastic Beral pipet. Invert the pipet and immerse the reservoir end of the pipet in the water bath. Allow both the test tube and the Beral pipet to remain in the water bath for at least two minutes before proceeding.

Repeat the collect pressure data steps mentioned above in Step 1. Remember to save and record the data.

4. Decompose 3% H_2O_2 solution with 0.5 mol • L^{-1} KI solution at 30 ℃

Conduct Step 4 identically to the procedure in Step 1, with one exception: set the water bath at 30 ℃.

DATA TREATMENT

1. Data table

	Reactants	Temperature (℃)	Initial rate (kPa • s^{-1})
1	4 mL 3.0% H_2O_2 +1 mL 0.5 mol • L^{-1} KI		
2	4 mL 3.0% H_2O_2 +1 mL 0.25 mol • L^{-1} KI		
3	4 mL 1.5% H_2O_2 +1 mL 0.5 mol • L^{-1} KI		
4	4 mL 3.0% H_2O_2 +1 mL 0.5 mol • L^{-1} KI		

2. Data analysis

	Initial rate $(mol \cdot L^{-1} \cdot s^{-1})$	$[H_2O_2]$ after mixing	$[I^-]$ after mixing	Rate constant k
1				
2				
3				
4				

3. Calculate the rate constant, k, and write the rate law expression for the catalyzed decomposition of hydrogen peroxide.

4. Use the Arrhenius equation (shown below) to determine the activation energy, E_a, for this reaction.

$$\ln \frac{k_1}{k_2} = \frac{E_a}{R} \left(\frac{1}{T_2} - \frac{1}{T_1} \right)$$

QUESTIONS

1. Does the initial concentration of hydrogen peroxide and KI affect the experimental results? How can we choose their concentrations?

2. The following mechanism has been proposed for this reaction:

$$H_2O_2 + I^- \longrightarrow IO^- + H_2O$$
$$H_2O_2 + IO^- \longrightarrow I^- + H_2O + O_2$$

If this mechanism is correct, which step must be the rate-determining step? Explain.

3. How to check whether the system leaks?

2.11 Paper Chromatography-Separation and Identification of Metal Cations

AIMS

1. To know the principle of the paper chromatography.

2. Known and unknown solutions of the metal ions will be analyzed using paper chromatography.

3. An unknown solution containing some cations will be identified by comparison to the R_f values and colors of the stained spots of known solutions.

INTRODUCTION

Most chemists and many other scientists must routinely separate mixtures and identify their components. The ability to qualitatively identify the substances found in a sample can be critical. Chromatography is one of the first tools used in such situations. The method depends on the different solubilities, or adsorptivities, of the substances to be separated relative to the two phases between which they are to be partitioned. In this technique, many types of mixtures can be separated into the component pure substances; by comparison to a standard sample, each component substance can also be tentatively identified.

Many varieties of chromatography exist, each one designed to separate specific types of mixtures. The common feature of each type of chromatography is that a mobile phase (a liquid or gas) is pushed through a stationary phase (a solid). Table 1 lists several varieties of chromatography and typical identities of the phases. Paper chromatography will be used in this experiment.

Table 2 – 4   **Varieties of chromatography**

Type of Chromatography	Mobile Phase	Stationary Phase
Gas (GC)	inert gas (helium)	waxy liquid or silicone inside narrow tubing
Liquid (LC, HPLC, column)	solvent/solvent Mixture (organic or aqueous)	solid packing (silica, alumina)
Paper	solvent/solvent Mixture (organic or aqueous)	paper
Thin-Layer (TLC)	solvent/solvent Mixture (organic or aqueous)	silica/alumina coated glass, plastic or metal

The example of column chromatography demonstrates the typical features found in this analytical technique. Figure 2 – 16 shows an experiment where a two-component mixture is subjected to column chromatography. A column is packed with a solid material called the stationary phase. A liquid solvent or eluting solution is poured into the column and completely wets the solid packing material. Then the mixture is loaded onto the top of the wet column and more eluent is added. Gravity pulls the mobile phase down through the stationary phase and the components in the mixture start to move through the column at different rates. Component A moves faster than component B; thus component B is retained on the column for a longer time than component A. Usually this is due to a difference in solubility of the two compounds in the solvent and/or to a difference in attraction to the solid packing material. As more eluent is added to the top of the column, the components will eventually exit the

column separately. The time taken to exit the
column, called retention time, will be
reproducible for each component under the given
set conditions—mobile and stationary phase
identities, temperature and column width. Once
the components exit the column, the solvent can
be removed by evaporation and the pure
components can be further analyzed or identified.

Tentative identification of the components
can be achieved by comparing the unknown
mixture a carefully prepared known mixture:
if a known component has the same retention
time as an unknown component under the same
conditions, it is probable, but not conclusive,
that the two components are the same.
Further analysis may be needed to confirm this

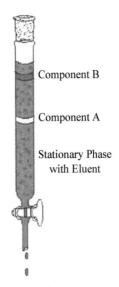

Figure 2 - 16 A typical column chromatography experiment demonstrates the separation of a two-component mixture.

hypothesis. If the known and the unknown have different retention times, then it is
not likely that the two components are identical.

In this experiment, similar principles are used to separate several metal cations
by a paper chromatography procedure. The metal ions: Fe^{3+}, Co^{2+}, Cu^{2+}, and Mn^{2+}
have differing solubility in the mobile phase (acetone with concentrated hydrochloric
acid and distilled water) and will move at different rates up the paper. The different
metal-ion solubilities are probably due to the formation of various compounds with the
chloride ion and their varying ability to dissolve in the organic solvent.

Figure 2 - 17 shows how to prepare the paper. Standard solutions containing each
of these ions will be spotted onto the paper using a capillary tube, along with a
standard solution containing all four ions separately and their mixture. Unknown
samples will also be spotted onto the paper. Once the paper is prepared, it will be
developed by placing the paper into the eluent. After 60 minutes, the paper is
visualized by wetting it with an aqueous solution containing potassium iodide, KI, and
potassium ferrocyanide, $K_4[Fe(CN)_6]$. The unique color observed for each ion is
produced by a chemical reaction with the visualization solution. This is one useful way
to identify which ions are present in an unknown mixture.

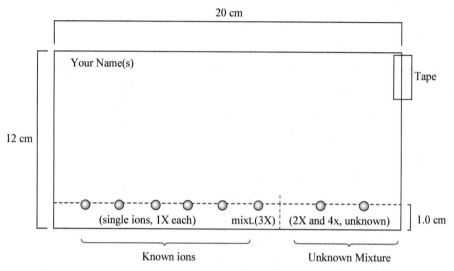

Figure 2 – 17   **How to prepare the paper for the chromatography experiment.**

The distance the ion moves up the paper can also be used to identify the ion. However, since students will develop their chromatography experiments for different amounts of time and under slightly different conditions, each student will have somewhat different measured distance for a given ion. The ratio of the distance moved by an ion (D) to the distance moved by the solvent (F, solvent front) is characteristic and should be nearly the same for all students. This ratio is called R_f, or "retention factor."

$$R_f = \frac{D}{F} \tag{2-11-1}$$

REAGENTS AND APPARATUS

- Filter paper, pencils, rulers, hair drier, scissors, glass capillary tubes, tape, 600 mL beakers, plastic wrap, paper towel.
- $0.1\ mol \cdot L^{-1}$ aqueous solutions of $Fe(NO_3)_3$, $Co(NO_3)_2$, $Cu(NO_3)_2$, and $Mn(NO_3)_2$, mixed solutions of the four ions, unkonown solutions, eluting solution (V : V : V (Acetone: conc. HCl : H_2O) = 19 : 4 : 2), visualizing solution ($0.5\ mol \cdot L^{-1}$ aqueous solution of KI and $K_4[Fe(CN)_6]$, volume ratio = 1 : 1).

PROCEDURES

1. Preparation of the paper for chromatography

Each student should obtain a piece of filter paper with the dimensions shown in Figure 3. Make sure the paper is clean and without tears or folds. Use a pencil—not a pen—and a ruler to draw a line across the paper one cm from the long edge of the paper. You will spot the metal ion solutions on this line. Write your name in pencil in the upper left-hand corner of the paper.

Practice spotting water and/or ion solutions onto a strip of filter paper so that you know how to create spots of the correct size. Use glass capillary tubes to spot the ions onto the paper. Solution is applied by lightly and quickly touching a capillary tube containing the solution to the line you drew on the paper. The spots should be between 5 – 8 mm in diameter. Spots larger than this will excessively spread out during the experiment and make analysis difficult.

Known 0. 1 M aqueous solutions of $Fe(NO_3)_3$, $Co(NO_3)_2$, $Cu(NO_3)_2$, and $Mn(NO_3)_2$ are provided in reagent bottles, each containing two or three capillary tubes. Starting on the left, mark the identity of the ion underneath each spot with a pencil; then spot each known ion carefully onto the line. Be careful to avoid contaminating the capillary tube with other ions and replace the capillary tubes back into the correct test tube. A reagent bottle containing a known mixture of all four ions is also provided with a set of capillary tubes. Spot this mixture onto the line as well. Because this solution is more dilute than the single-ion known solutions, apply the known mixture three times, letting the spot dry between each application. A hair drier will help to dry the spot more quickly.

Several unknowns are also provided in reagent bottles, along with capillary tubes. Your instructor will tell you which unknown should be used. The unknowns will contain between one and four cations, and are more dilute than the single-ion known solutions. The unknown will also need to be applied two and four times for the two trials, letting the spot dry between each application.

2. Developing the chromatography paper

Place a piece of tape along the upper right edge, as shown in Figure 2 – 17 Then form a cylinder by connecting the two short edges of the paper with the tape. Make sure the edges do not touch. The paper should look similar to Figure 2 – 18.

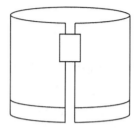

**Figure 2 − 18   Folded paper should look like this prior
to developing the experiment.**

Obtain 15 mL of the eluting solution. Carefully pour some of this solvent into a 600 mL beaker and carefully swirl it for a second or two. Caution: Do not breathe the vapors from this solution! Make sure that the level of the liquid will be below the spot line on the paper once the paper is placed in the developing chamber.

Place the paper cylinder into the beaker with the marked edge down. The spots should be above the level of the solvent. The paper should not touch the sides of the beaker. Carefully cover the beaker with plastic wrap and place it in the hood for about 60 minutes. The solvent should start to move up the paper. Once the beaker is covered, make sure it is level and does not disturb it during the development period. Your instructor may have an assignment for you to work on while you wait.

3. Visualization and analysis of the paper

Once the development period is over, wear disposable gloves and remove the paper from the beaker. Let any solvent drip back into the beaker, then remove the tape. Lay the chromatography paper on a paper towel and immediately mark the solvent front with a pencil. Pour the used eluting solvent into the waste container provided. Dry the paper with a hair drier in the hood. Caution: Do not breathe the vapors!

Once the paper is dry, bring it to the visualization station on the paper towel. Briefly dip the paper into the visualizing solution located in a shallow dish in the fume hood. Lift the paper out of the solution immediately and let any excess drip off at the station. Place the wet paper onto a dry paper towel and dry it with a hair drier immediately, then carry it to your bench for analysis.

Find each known single-ion first and record the colors you observe. Some spots may fade over time, so record the colors while the paper is still wet. Measure the distance each spot moved, D, with a ruler. Measure to the center of each spot. Record your data in the data table.

Measure the distance to the solvent front, F. The value of F should be approximately the same across the entire paper. Use these values to calculate the R_f for each ion. Make your measurements as shown in Figure 2 – 19. Each observed spot has its own R_f value. Record your results in the data table.

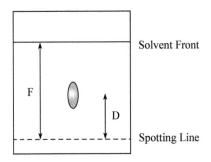

Figure 2 – 19   Measurement of distances used in the calculation of R_f for a spot.

In the lane containing the mixture, find each ion and record the distance moved by each ion. Calculate the R_f for each ion in this lane. The values should closely match those observed in the single-ion knowns.

In the lane containing the unknowns, locate the center of each spot observed and record its distance and calculate the R_f values. Use the lane that has the clearest spots. The color and R_f values for the unknown spots should closely match some of the known ions. You should now be able to identify which ion or ions are found in your unknown. Record your data in the corresponding table.

Make a sketch of your chromatogram in the space provided on your lab report form, being sure to indicate the position and approximate size and shape of each spot on the paper. Dispose of the paper in the designated waste container.

4. Clean up

Place the chromatography paper and the used gloves in the waste container provided. The used eluting solution should already have been placed into another waste container. Note that two different waste containers are provided for this experiment so be sure to read the labels so you will use the correct one! Be sure to wash your hands thoroughly before leaving the laboratory.

DATA TREATMENT

1. Sketch of Chromatogram.
2. Data for known ions.

Ion	Spot Color (Stained)	D (Single-Ion)	F (Single-Ion)	R_f	D (Ion Mixt.)	F (Ion Mixt.)	R_f
$Fe(NO_3)_3$							
$Co(NO_3)_2$							
$Cu(NO_3)_2$							
$Mn(NO_3)_2$							

3. Data for unknownions.

Spot Number (from lowest D)	Spot Color (Stained)	D (Unknown)	F (Unknown)	R_f	Identity of Spot
1					
2					
3					
4					

QUESTIONS

1. What criteria were used to identify the ion(s) found in your unknown? Include any difficulties in identifying any ions.

2. If you let the experiment run for only 10 minutes, what would be the likely result? Would any problems arise in identification of the unknown?

3. If Co^{2+} and Cu^{2+} spots were the same color, would the identification of an unknown be any more difficult?

2.12 Extraction of Spinach Pigments and Thin Layer Chromatography

AIMS

1. To learn the method of extracting chlorophyll from leaves.

2. To understand the principle of TLC and master the general operation and qualitative identification methods of TLC.

3. To master the quantitative method of spectrophotometry.

INTRODUCTION

The leaves of plants contain a number of colored pigments generally falling into two categories, chlorophylls and carotenoids. The concentrations of pigments are different for different plant species and also depend on the time of year. Also, these pigments have different solubility in different solvents. The green chlorophylls a and b, which are highly conjugated compounds capture the (nongreen) light energy used in photosynthesis. Orange carotenoids are part of a larger collection of plant-derived compounds called terpenes.

Chlorophyll is a green pigment found in most plants, algae, and cyanobacteria. Chlorophyll absorbs light most strongly in the blue portion of the electromagnetic spectrum, followed by the red portion. Conversely, it is a poor absorber of green and near-green portions of the spectrum, which it reflects, producing the green color of chlorophyll-containing tissues.

Chlorophyll molecules are specifically arranged in and around photosystems that are embedded in the thylakoid membranes of chloroplasts. Two types of chlorophyll exist in the photosystems of green plants: chlorophyll a and b. Chlorophyll a is a blue-black plant pigment having a blue-green alcohol solution; found in all higher plants. Chlorophyll b is a dark-green plant pigment having a brilliant green alcohol solution; generally characteristic of higher plants. Chlorophyll is a chlorin pigment, which is structurally similar to and produced through the same metabolic pathway as other porphyrin pigments such as heme. At the center of the chlorin ring is a magnesium ion. For the structures depicted in this article, some of the ligands attached to the Mg^{2+} center are omitted for clarity. The chlorin ring can have several different side chains, usually including a long phytol chain. There are a few different forms that occur naturally, but the most widely distributed form in terrestrial plants is chlorophyll a.

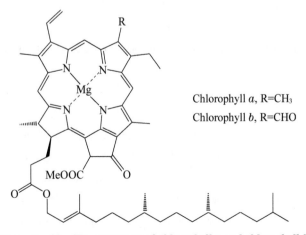

Chlorophyll a, R=CH$_3$
Chlorophyll b, R=CHO

Figure 2 – 20   The structures of chlorophyll a and chlorophyll b

Carotenoids are yellow, orange, or red pigments (cyclic or acyclic isoprenoids) synthesized by bacteria, fungi, and higher plants. The carotenoids include carotene and xanthophylls, which are widely found in plants; Lycopene ($C_{40}H_{56}$), found in the fruits of the tomato, dog rose, and nightshade; Zeaxanthin ($C_{40}H_{56}O_2$), found in corn kernels; Vi-olaxanthin and flavoxanthin, found in squashes and gourds; Cryptoxanthin ($C_{40}H_{56}O$), found in papaya; Physalin ($C_{72}H_{116}O_4$), found in the flowers and fruits of Physalis; Fucoxanthin ($C_{40}H_{56}O_6$), found in brown algae; Crocetin ($C_{20}H_{24}O_4$), found in the stigmata of saffron; and Taraxanthin ($C_{40}H_{56}O_4$), found in the flowers of snapdragon and coltsfoot. The relative content of the various carotenoids changes in the course of development of the plant and under the influence of environmental conditions. The concentration of carotenoids is highest in the plastids of the cells.

Carotenoids promote the fertilization of plants by stimulating the germination of pollen and the growth of the pollen tubes. They play a part in the absorption of light by plants and in the perception of light by animals. They are also a major factor in the processes of photosynthesis and oxygen transport in plants. The number and position of the double bonds in the molecules of the carotenoids determine their color; over 150 carotenoids (pigments) are known. Carotenoids with a larger number of double bonds absorb in the long-wave part of the spectrum, and their color is bright orange or red.

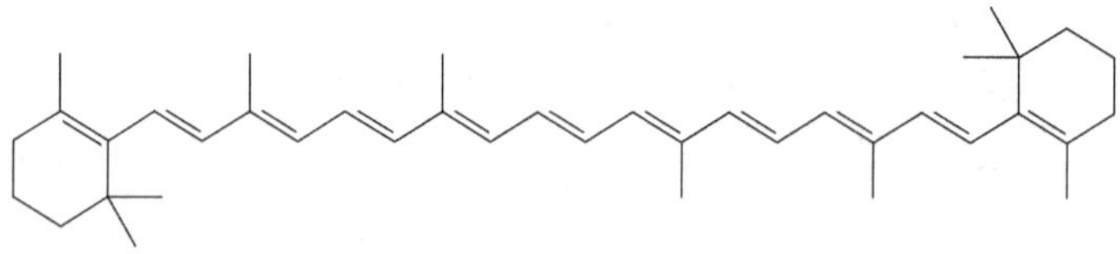

beta-carotene

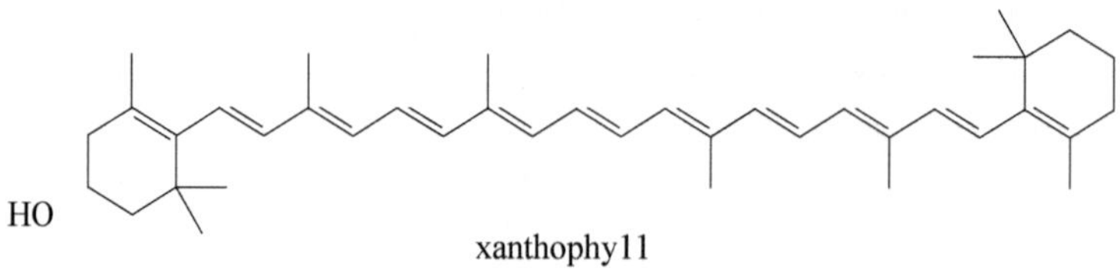

xanthophyll

Figure 2 - 21   The structures of beta-carotene and xanthophyll

Chlorophyll a and b are complexes of pyrrole derivatives with magnesium metal. Although they contain some polar groups, their large alkyl structure makes them soluble in organic solvents such as acetone, ethanol, ether and petroleum ether. Beta-carotene and xanthophyll are lipid-soluble tetraterpenoids. Compared with carotene, xanthophyll is soluble in alcohol and less soluble in petroleum ether. According to their solubility in organic solvents, they can be extracted from plant leaves. Plant chloroplast pigments are usually extracted with acetone, ethanol, ether, acetone-ether, methanol-petroleum ether and other organic solvents.

Thin-layer chromatography (TLC) is more unique in separating and identifying natural products. It is defined as the separation of a mixture of two or more different compounds by distribution between two phases, one of which is stationary and the other moving. In thin layer chromatography (TLC) the "stationary" phase is the adsorbent silica, which is bound to an aluminum-backed plate (also called a TLC plate). Silica is considered a polar substance since the surface of the crystals consists of polar hydroxyl (OH) groups. The "moving" phase is an organic solvent system that, by capillary action, will move up the stationary silica coated plate. All solvent systems will be considered non-polar relative to the silica adsorbent.

The sample mixture is usually applied as a small spot near the base of the TLC plate (called "spotting"). The plate is then put into a solvent reservoir where, by capillary action, the solvent will rise up the plate. As the solvent ascends the plate, the compounds in the sample are partitioned between the moving liquid solvent and the stationary solid phase. This process is called developing the TLC plate.

When developing a TLC plate, the various components in the mixture are separated. This separation is based upon each compound's distribution equilibrium between the solventand adsorbent.

Each compound will have a unique distribution equilibrium depending mainly upon the polarity of the compound (based on intermolecular forces between the compounds being separated and the adsorbent). An example is that of a 'polar' compound vs. a non-polar compound. The distribution equilibrium of a polar compound will favor the adsorbent since the adsorbent is highly polar ("like dissolves/attracts like"). The 'non-polar' compound however, will have less affinity for the polar adsorbent and will have an equilibrium favoring solubility in the mobile solvent. The consequence of this is that polar compounds will 'stick' to the stationary TLC plate while non-polar compounds will separate and travel upward with the solvent. When developing a TLC plate we can state that each compound in the mixture will ascend the plate at a different rate; polar compounds ascend slowly, less polar compounds ascend quickly.

Spinach leaves, which the students will use in this experiment, contain chlorophyll a, chlorophyll b and carotene as major pigments as well as smaller amounts of other pigments such as xanthophylls. In this experiment the student will isolate the spinach pigments using differences in polarity to effect the separation. The extract of plant pigments will be further separated into chlorophyll a, chlorophyll b, beta-carotene and xanthophyll by TLC. Since the different components are colored differently, the separation is easily followed visually.

REAGENTS AND APPARATUS

- Analytical balance, scissors, mortar and pestle, centrifuge, matched centrifuge tubes, Pasteur pipettes, test tubes, TLC plates, micro capillaries, pencil, ruler, 600 mL beakers, 5 mL graduated cylinder, plastic wrap.
- Spinach leaves, anhydrous sodium sulfate (Na_2SO_4), acetone, hexane, distilled water.

PROCEDURES

1. Isolation of pigment from spinach leaves

Weigh about 1. 0 g of fresh spinach leaves (avoid using stems or thick veins). Cut or tear the spinach leaves into small pieces and place them in a mortar along with 0. 5 g of anhydrous sodium sulfate (Na_2SO_4) and 2. 0 mL of acetone. Grind with a pestle until the spinach leaves have been broken into particles too small to be seen clearly and mix well with Na_2SO_4.

If too much acetone has evaporated, you may need to add an additional portion of acetone (0. 5—1. 0 mL). Using a Pasteur pipette or spatula, transfer the mixture to a centrifuge tube. Rinse the mortar and pestle with 2. 0 mL of cold acetone, and transfer theremaining mixture to the centrifuge tube. Cap tightly. Centrifuge the mixture being sure to balance the centrifuge first (Make sure to return the centrifuge tube to the TA at the end of the experiment.).

Add ~2. 0 mL of hexane to the centrifuge tube, cap the tube, and shake the mixture thoroughly. Next, add 2. 0 mL of water and shake thoroughly with occasional venting. Centrifuge the mixture to break the emulsion, which usually appears as a cloudy green layer in the middle of the mixture. The pigment layer is the top layer, which should be dark green. Most of the acetone will dissolve in the water.

Using a dry Pasteur pipette, carefully separate layers and transfer the top organic layer (a dark-green hexane solution of spinach pigments) into a clean test tube. Add another 1. 0 mL of hexane to the centrifuge tube that contains the aqueous layer, cap the tube and centrifuge the mixture to break the emulsion. Separate the layers again and add the top layer to the test tube.

The dark-green hexane solution of spinach pigments in the test tube may contain traces of water that must be removed before separating the components through chromatography. To dry the solution, add 0. 5 g of anhydrous sodium sulfate (Na_2SO_4) to the hexane solution. Cap and gently swirl to allow the sodium sulfate to contact all parts of the hexane. After standing for 5 minutes, use a clean, dry Pasteur pipette to transfer the liquid into another clean test tube. Label this test tube with an E for extract so that you do not confuse it with the test tubes you will be working with later in this experiment.

Add about 0. 5 mL hexane to rinse the hydrated sodium sulfate and transfer this liquid to the same test tube.

2. TLC of spinach extract

Obtain two 1-inch TLC plates and micro capillaries from your instructor. These plates have a flexible backing, but they should not be bent excessively. They should be handled carefully or the adsorbent may flake off them. Also, they should be handled only by the edges; the surface should not be touched. Using a lead pencil (not a pen) lightly draw a line across the plates (short dimension) about 1 cm from the bottom (on the coated side). At the center of this line make a light mark. This is the point at which the spinach extract will be spotted.

To spot the TLC plate, fill the capillary tube by dipping one end into the spinach extract. Capillary action fills the pipet. Empty the pipet by touching it lightly to the thin-layer plate at the mark that is at the center of the 1 cm line from the bottom (The spot must be high enough so that it does not dissolve in the developing solvent). When the pipet touches the plate, solution is transferred to the plate as a small spot. The spot should be no larger then 2 mm in diameter and should be a fairly dark green. If you do not have a dark green spot, you may spot again using another sample of your spinach extract. Allow the solvent to evaporate completely between successive applications, and spot the plate in exactly the same position each time. It is important that the spots be made as small as possible and that the plates not be overloaded.

When the first plate has been spotted it is ready to be placed in a development chamber. For a development chamber you will use your large beaker lined with a piece of filter paper and cover (see Figure 2 – 22).

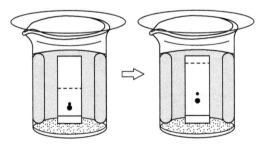

Figure 2 – 22   TLC Development Chamber

When the development chamber has been prepared, obtain a small amount of the development solvent (70/30 mixture of hexane/acetone). Fill the chamber with the development solvent to a depth of about 1/4 inch (about 10 mL of solvent). Be sure that the liner is saturated with the solvent. The solvent level must not be above the spots on the plate or the samples will dissolve off the plate into the reservoir instead of developing. Place the spotted plate in the chamber and allow the plate to develop (the solvent will slowly move up the TLC plates). Since the backing on the TLC plates is very thin, if they touch the filter paper liner of the development chamber at any point, solvent will begin to diffuse onto the adsorbent surface at that point.

When the solvent has risen to a level about 1 cm from the top of the plate, remove the plate from the chamber and, using a lead pencil, mark the position of the solvent front. Let the plate dry. Lightly outline all the observed spots with a pencil. Before proceeding, make a sketch of the plate in your notebook and label each spot by color. Using a ruler marked in millimeters, measure the distance that each spot has traveled relative to the solvent front. Under an established set of such conditions, a given compound always travels a fixed distance relative to the distance of the solvent front. This ratio of the distance the compound travels to the distance the solvent travels is called the R_f value. This can be expressed as a decimal fraction:

R_f=distance traveled by substance/distance traveled by solvent front (see Figure 2 – 23). Calculate R_f values for each observed spot.

Repeat the above TLC of you spinach extract but use a different proportion of hexane/acetone for your developing solvent (be sure to record the proportion used in your notebook). Once this plate has developed, sketch the plate in your notebook, label the spots by color and calculate R_f values for all the spots as described above.

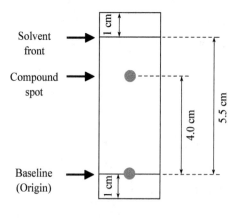

$$R_f = \frac{4}{5.5} = 0.73$$

Figure 2 – 23 R_f Values

Waste Disposal: Put all extra acetone and hexane solutions into the organic waste. Aqueous solutions can go down the drain. Rinse centrifuge tubes with water and place them into the dirty glassware bin. Rinse the mortar and pestle with a small amount of acetone (put rinse into organic waste), rinse with water and put into the dirty glassware bin. Spinach residue can go into the trash. Pasteur pipets should go into the glass waste container (unless otherwise instructed).

DATA TREATMENT

1. The original TLC.
2. Data from TLC.

	Spot color	Spot distance	Solvent front	R_f	Type of pigment
1					
2					
3					
4					

QUESTIONS

1. Why traces of water in the dark-green hexane solution in the test tube must be removed before separating the components through chromatography?

2. The TLC must be air-dried before drawing fine lines with pencil. Why not dry it with fire or in an oven?

3. What are the pigments in the green leaves, and what colors are they? Do they have the same content in the green leaves?

2.13 Purification on Raw Salt

AIMS

1. To master the principle and method of purification on raw salt.

2. Practice on the operations of dissolution, precipitation, filtration, evaporation, crystallization, drying and so on.

3. To learn how to qualitatively identify SO_4^{2-}, Ca^{2+}, and Mg^{2+} ions.

INTRODUCTION

Recently, the annual world production of salt has exceeded 250 million tons. Approximately one third of the total is produced by solar evaporation of sea water or inland brines. Another third is obtained by mining of rock salt deposits, both underground and on the surface. The balance is obtained as brines, mainly by solution mining. Brines can be used directly (for example in diaphragm electrolysis) or thermally evaporated to produce vacuum salt.

Salt type	World production
Solar salt	90,000,000 t/y
Rock salt	80,000,000 t/y
Brines	80,000,000 t/y
Total	250,000,000 t/y

The purity of washed solar salt produced in India and China reach 99%—99.5% (NaCl, dry base) but solar salt produced in Australia and Mexico is 99.7%—99.8% pure. The purity of processed rock salt fluctuates between 97% and $99\%+$ in the USA and in Europe. Vacuum salt is usually 99.8%—99.95% pure.

The chemical industry is the largest salt consumer of salt using about 60% of the total production. This industry converts the salt mainly into chlorine, caustic soda and soda ash without which petroleum refining, petrochemistry, organic synthesis, glass production, etc. would be unthinkable.

Salt user	Salt consumption
Chemicalindustry	60%
Food	30%
Other	10%

The second largest user of salt is mankind itself. Humans need about 30% of the total salt produced to support their physiological functions and eating habits. Salt for food is the most "taken for granted" commodity, available from thousands of sources in hundreds of qualities as table, cooking and salt for food production. About 10% of salt is needed for road de-icing, water treatment, production of cooling brines and many other, smaller applications. Whatever the use of salt, it is the sodium chloride in the salt that is required and not the impurities. The purer the salt, the more valuable it is.

Sodium chloride in salt is always the same. It is the "non-salt" in salt-the impurities-that make the difference. In fact, the multiplicity of impurities in salt and their relative quantities are so variable that every salt needs to be considered on its own merits.

Except for insolubles, the origin of impurities is the sea water. Solar sea salts, as a rule just few months old, are rather similar. Rock salts, millions of years old, may vary greatly, from pure to dirty, from white to black. Lake salts contain components leached from the ground of the surrounding rocks in variable quantities. Salt lake chemistry is a science of its own.

Calcium sulphate is the most persistent companion of salt. In rock salt, calcium sulphate is found as anhydrite, hemihydrite or polyhalite. Gypsum is found both in sea salt and in lake salt. Natural brines are, as a rule, saturated with calcium sulphate.

Magnesium salts are always present in the sea salt, usually at a ratio of approx. One and a half weight units of magnesium chloride to one weight unit of magnesium sulphate. In lake salts, magnesium sulphate is usually accompanied by sodium sulphate, for example in Sambhar Lake salts from Rajasthan in India or in Azraq salts from Jordan. Magnesium chloride also occurs together with calcium chloride, for example in the Dead Sea brines where also potassium chloride and sodium bromide are found in exceptionally high concentrations. Insolubles are present in salts of all origins in greatly fluctuating quantities.

	Rock salt	Sea salt	Lake salts	Brines
$CaSO_4$	0.5%—2%	0.5%—1%	0.5%—2%	saturated
$MgSO_4$	Traces	0.2%—0.6 %	Traces	Traces
$MgCl_2$		0.3%—1%	Traces	
$CaCl_2$			Traces	
Na_2SO_4			Traces	
KCl			Traces	
NaBr			Traces	
Insolubles	1%—10%	0.1%—1%	1%—10%	

In the chemical industry, salt is mostly dissolved together with the impurities in water or brine. Prior to feeding to the process, the brine is purified. Failure to purify the brine may have serious, even lethal consequences.

In electrolytic cells, excessive magnesium causes hydrogen evolution on the anode. Hydrogen and chlorine form an explosive mixture. Explosion in the cells or in the chlorine liquefaction may damage the equipment and release chlorine to the environment. Chlorine gas is highly poisonous. Stringent safety measures are taken in the chloralkali industry to avoid this happening. The elimination of magnesium is of prime concern.

Impure brine in mercury cells will cause butter formation. Butter will disturb mercury flow, causing short circuits that burn the electrodes. Alternatively, a large electrode gap must be maintained which will increase the power consumption. Butter removal will expose workers to mercury vapours that are damaging to health. Disposal of mercury butter is costly and undesirable for the environment.

Sludge from brine purification in chloralkali plants with mercury cells is contaminated with mercury. Sludge decontamination by distillation requires high temperatures, is costly and never complete. The disposal of mercury contaminated sludge is environmentally objectionable and very costly. Avoiding the formation of sludge is better than having to dispose of it. This requires salt of high purity.

Calcium and magnesium will damage the ion exchange membranes irreversibly. Erratic impurity content in salt may cause hardness breakthrough to the membrane cells. Membranes cost a fortune. Prices of USD 600—1 000/m^2 have been reported. The purer the salt, the more remote is the danger of membrane damage.

In soda ash production, excessive sulphate reduces the value of the product. Accumulating calcium in the process causes encrustations. Periodical scale removal is costly and leads to loss of production. Salt may be a cheap commodity. But impurities in salt and their removal cost in many cases more than the salt itself.

In the chemical industry, impurities in brine such as calcium and magnesium areprecipitated with chemicals. Sulphates are removed either by precipitation with barium or calcium or are controlled by purging the brine. In this experiment, crude salt will be purified by similar methods in the chemical industry.

REAGENTS AND APPARATUS

- Analytical balance, oven, heating mantles, water pump, 250 mL beakers, 10 mL and 100 mL graduated cylinders, brinell funnels, 500 mL filter bottles, droppers, glass rods, filter paper, adapters
- Raw salt, 6.0 mol $\cdot$ L^{-1} concentrated hydrochloric acid (HCl), 1.0 mol $\cdot$ L^{-1} barium chloride (BaCl$_2$) solution, saturated sodium carbonate (NaCO$_3$) solution, 2 mol $\cdot$ L^{-1} acetic acid (HAc), saturated ammonium oxalate ((NH$_4$)C$_2$O$_4$) solution, 6 mol $\cdot$ L^{-1} sodium hydroxide (NaOH) solution, 10 ppm magnesium reagent (azoviolet) solution, distilled and deionized water

PROCEDURES

1. Dissolution of raw salt

Weighing about 20 g raw salt in 250 ml beaker and record the mass to 4 decimal places. Add 80 mL water, and then put the beaker in the heating mantle. Heating and stirring till near boiling, to dissolve raw salt. Insoluble impurities sink to the beaker bottom and the total volume should be maintained at 70—80 mL.

2. Removal of SO_4^{2-}

Maintain the near boiling state, and then add 3—5 mL 1.0 mol · L^{-1} $BaCl_2$ solution while stirring. Continue boiling for 5 minutes, and keep the total volume unchanged. This heating operation will help the precipitated particles to grow up and easy to settle.

Remove the beaker from the heating mantle. After sedimentation, add 1—2 drops of 1.0 mol · L^{-1} $BaCl_2$ solution to the supernatant. If turbidity occurs, it means that SO_4^{2-} is not removed. Continue to add 1.0 mol · L^{-1} $BaCl_2$ solution to remove the remaining SO_4^{2-}.

If it is not turbid, it means that SO_4^{2-} has been removed. After cooling, filter with water pump, wash the precipitate with a small amount of distilled water three times, and keep the mother liquor.

3. Removal of Ca^{2+}, Mg^{2+} and Ba^{2+}

The filtrate is heated to near boiling and then add the saturated Na_2CO_3 solution drop by drop while stirring until no precipitation occurs. Add 0.5 mL Na_2CO_3 solution, and settle aside.

After sedimentation, add 1—2 drops of 1.0 mol · L^{-1} Na_2CO_3 solution to the supernatant. If turbidity occurs, it means that Ba^{2+} has not been removed, and the saturated Na_2CO_3 solution should be added to the original solution until Ba^{2+} is removed.

If it is not turbid, it means that Ba^{2+} has been removed. After cooling, filter with water pump, wash the precipitate with a small amount of distilled water three times, and keep the mother liquor.

4. Removal of excess CO_3^{2-}

Drop 6.0 mol · L^{-1} HCl into the filtrate, heat and stir, and neutralize the pH of the solution to 2—3 (check with pH test paper).

5. Concentration and Crystallization

The solution then is concentrated until a large amount of NaCl crystals appeared (about 1/4 of the original volume). The solution is cooled to room temperature and filtered with water pump. Then wash the crystals with a very small amount of distilled water and drained.

The NaCl crystals are then transferred to an evaporating dish and dried at 100 ℃ for 1 hour. Take out the NaCl crystals with a baking glove. Be careful, it's very hot. After cooling, weigh and calculate the yield.

6. Testing of product purity

Take 1 g of the raw salt and the purified product, respectively, dissolved in 5 mL distilled water, and then carry out qualitative test of the following ions:

a. SO_4^{2-}: Take 1 mL of the two prepared solution, each in one test tube. Each tube is added 2 drops of 6.0 mol・L^{-1} HCl solution and 2 drops of 1.0 mol・L^{-1} $BaCl_2$ solution respectively, compare the precipitation in the two test tubes.

b. Ca^{2+}: Take 1 mL of the two prepared solutions, each in one test tube. Each tube is added 1—2 drops of 2 mol・L^{-1} HAc to acidify, and then added 3—4 drops of saturated $(NH_4)C_2O_4$ solution respectively. If white CaC_2O_4 precipitates, it indicates the presence of Ca^{2+}. Compare the precipitation in the two test tubes.

c. Mg^{2+}: Take 1 mL of the two prepared solutions, each in one test tube. Each tube is added 5 drops of 6.0 mol・L^{-1} NaOH and then added 2 drops of magnesium reagent, respectively. If a sky blue precipitate is formed, it indicates the presence of Mg^{2+}. Compare the precipitation in the two test tubes.

DATA TREATMENT

1. Purification of raw salt

	Before purification	After purification	Yield
NaCl			

2. Testing of product purity

test	methods		results
	1 mL raw solution	1 mL product solution	comparison
SO_4^{2-}	① 2 drop of 6.0 mol/L HCl solution ② 2 drop of 1.0 mol/L $BaCl_2$ solution		
Ca^{2+}	① add 2.0 mol/L HAc to pH<7 ② 3—4 drop of saturated $(NH_4)_2C_2O_4$ solution		
Mg^{2+}	① 5 drop of 6 mol/L NaOH solution ② 2 drop ofmagnesium reagent		

QUESTIONS

1. Can we remove the Ca^{2+} and Mg^{2+} firstly before SO_4^{2-} in the experiment? Why?

2. According to the use of 6.0 mol $\cdot$ L^{-1} HCl solution in the experiment, please make sure the procedure for preparing 6.0 mol $\cdot$ L^{-1} from concentrated HCl. (Assumed that the volume of 6 mol $\cdot$ L^{-1} HCl is 1 000 ml, identify the devices you need, and calculate the volume of concentrated HCl.)

2.14 Some Properties on Alkali Metals and Halogens

AIMS

1. To learn the origin of spectral emission in the visible region and more fully appreciate the lectures on atomic structure.

2. To master the factors which affect the solubility of a salt.

3. The importance of relative ionic radii in determining lattice enthalpy.

4. Redox reactions and predicting reactivity.

INTRODUCTION

Group 1A chemistry

The chemistry of the alkali metals is almostentirely limited to oxidation state+1.

This is not because loss of an electron by a gaseous alkali metal atom is an exothermic process, it is not,

$$\text{e. g. } Na(g) = Na^+(g) + e^-, \Delta H \approx 500 \text{ kJ} \cdot \text{mol}^{-1} \quad (2\text{-}14\text{-}1)$$

but rather reflects the lattice energies or solvation energies of the compounds formed by the alkali metals.

$$\text{e. g. } Na^+(g) + Cl^-(g) = NaCl(s), \Delta H \approx -900 \text{ kJ} \cdot \text{mol}^{-1} \quad (2\text{-}14\text{-}2)$$

The tests in this experiment illustrate some of the differences in properties of alkali metal compounds. Most of these differences can be attributed directly or indirectly to the different relative sizes of the unpositive cations or their solvated ions.

For Group 1A compounds, flame tests are usually by far the easiest way of identifying which metal you have got. For other metals, there are usually other easy methods which are more reliable, but the flame test can give a useful hint as to where to look. In a flame test the colours arise from the excitation of electrons in an atom and emssion of light as the electrons drop to a lower energy level.

Table 2 - 5

	flame colour
Li	red
Na	strong persistent yellow-orange
K	lilac (pink)
Rb	red (reddish-violet)
Cs	blue-violet
Ca	orange-red
Sr	red
Ba	pale green
Cu	blue-green (often with white flashes)
Pb	greyish-white

Group 7A chemistry and redox behaviour

It is a characteristic of Group 7A elements, halogens, that they can show a range of different oxidation states. This is in contrast to Group 1A elements. One stable oxidation state is -1, but the elements Cl, Br, and I have stable compounds in a variety of different oxidation states.

The stability of the different oxidation states varies as you descend the Group. In order to predict the products of the redox reactions you will need to be able to balance redox equations.

This is are a useful and easy way of deducing the overall stoichiometry of a redox reaction. For example, consider the oxidation of chromium (II) chloride by potassium dichromate in acid solution. The product of the reaction is chromium (III). Firstly, we write a half equation for the oxidation of Cr(II) to Cr(III):

$$\text{e. g. } Cr^{2+} = Cr^{3+} + e^- \tag{2-14-3}$$

Secondly, we write a half equation for the reduction of $Cr_2O_7^{2-}$ [Cr(VI) to Cr(III)]:

$$\text{e. g. } Cr_2O_7^{2-} + 14H^+ + 6e^- = 2Cr^{3+} + 7H_2O \tag{2-14-4}$$

NB: (i) The half equations must balance. They are balanced electronically by the inclusion of the appropriate number of electrons. (ii) If both the oxidant and reductant contain a common element it will generally end up in the same oxidation state. (iii) Counter ions which take no part in the reaction are omitted.

Finally, to obtain the overall equation we combine (2-14-1) and (2-14-2) and eliminate the electrons. In this case we multiply (2-14-1) by 6 and add the result to (2-14-2):

$$6Cr^{2+} + Cr_2O_7^{2-} + 14H^+ = 8Cr^{3+} + 7H_2O \tag{2-14-5}$$

REAGENTS AND APPARATUS

- Tubes, 100 mL graduated cylinder, droppers
- Lithium chloride (LiCl), sodium chloride (NaCl), potassium chloride (KCl), 0.1 mol $\cdot$ L^{-1} potassium fluoride (KF) solution, 0.1 mol $\cdot$ L^{-1} potassium carbonate (K$_2$CO$_3$) solution, 0.1 mol $\cdot$ L^{-1} sodium perchlorate (NaClO$_4$) solution, ethanol, 0.01 mol $\cdot$ L^{-1} chlorine (Cl$_2$) water, 0.01 mol $\cdot$ L^{-1} bromine (Br$_2$) water, 0.01 mol $\cdot$ L^{-1} iodine (I$_2$) solution, 0.1 mol $\cdot$ L^{-1} sulphuric acid (H$_2$SO$_4$), 0.1 mol $\cdot$ L^{-1} sodium chloride (NaCl) solution, 0.1 mol $\cdot$ L^{-1} sodium bromide (NaBr) solution, 0.1 mol $\cdot$ L^{-1} sodium iodide (NaI) solution, hexane, 0.1 mol $\cdot$ L^{-1} sodium chlorate (NaClO$_3$) solution, 0.1 mol $\cdot$ L^{-1} sodium bromate (NaBrO$_3$) solution, 0.1 mol $\cdot$ L^{-1} sodium iodate (NaIO$_3$) solution, 6 mol $\cdot$ L^{-1} and 0.1 mol $\cdot$ L^{-1} hydrochloric acid (HCl), 0.1 mol $\cdot$ L^{-1} sodium thiosulphate (Na$_2$S$_2$O$_3$) solution, 0.1% methyl orange solution

PROCEDURES

1. The properties of lithium, sodium and potassium salts

① Label nine test tubes 1 to 9. To tubes 1—3 add enough LiCl to cover the end of a dry spatula. Add approximately 1 mL of water. Shake each tube. Repeat using NaCl in tubes 4—6 and KCl in tubes 7—9.

② To each of tubes 1, 4 and 7 add about 2 mL of the KF solution provided.

③ To each of tubes 2, 5 and 8 add about 2 mL of K_2CO_3 solution. Note any immediate change. If none, gently warm the solutions (to about 60 ℃).

④ To each of tubes 3, 6 and 9 add about 2 mL of $NaClO_4$ solution.

⑤ Place a small amount of solid LiCl in a test-tube and add about 5 mL of ethanol. Stir or shake for a couple of minutes. Repeat using NaCl and KCl.

2. Colour of elemental halogens

Take 3 clean, dry test tubes. Add about 1 mL aqueous halogen solution to each test tube and note the colour. Then add an equal volume of hexane to each. Shake the mixture well and allow to settle. Note down the colours of the two layers.

3. Formation of elemental halogens

(a) Take three test tubes and label them 1, 2, and 3. Carry out the following tests:

① test tube, add 1 mL dil. H_2SO_4, 5 drops NaCl, 5 drops $NaClO_3$.

② test tube, add 1 mL dil. H_2SO_4, 5 drops NaBr, 5 drops $NaBrO_3$.

③ test tube, add 1 mL dil. H_2SO_4, 5 drops NaI, 5 drops $NaIO_3$.

(b) Add 3—5 mL hexane to each of the tubes (Nos. 1—3) and shake well.

4. Formation of higher oxidation states of halogens

Take one test tube, add 5 drops $NaIO_3$, 2 drops NaI, 3 mL conc. HCl. Stir well. Note your observations and deductions. Add 1 mL hexane. Stir again. Note your observations and deductions. Add 10 drops NaI. Stir well. Note your observations and deductions.

5. The role of protons in the oxidation of iodide by iodate

Take one test tube, add 5 drops NaI, 5 drops $NaIO_3$, 5 mL $Na_2S_2O_3$ (sodium thiosulphate), 1 drop methyl orange. Dropwise with stirring add dil. HCl. Note your observations and deductions. (Hint: how many drops of HCl added before the indicator changed colour? —you may want to test dil HCl alone with methyl orange indicator to see how many drops required to change colour to acidic).

6. The relative stabilities of the halogens in their − 1 and 0 oxidation states

① With six test tubes labelled 1 to 6 carry out the following tests using 1 mL of the sodium halide solution in each case and record your initial observations.

Test tube		Add
1	NaCl soln	2 drops Br_2 soln
2	NaCl soln	2 drops I_2 soln
3	NaBr soln	10 drops Cl_2 soln
4	NaBr soln	2 drops I_2 soln
5	NaI soln	10 drops Cl_2 soln
6	NaI soln	2 drops Br_2 soln

② Add 1 mL of hexane to each of the test tubes (nos. 6 − 11). Shake well. Record the colours of the layers.

DATA TREATMENT

1. The properties of lithium, sodium and potassium salts

Test	①	②	③	④	⑤
Reagent	Water	KF soln	K_2CO_3 soln	$NaClO_4$ soln	ethanol
LiCl					
NaCl					
KCl					

2. Colour of elemental halogens

	Colour before adding hexane	Colour of lower layer	Colour of upper layer
Aqueous or organic?	Aqueous		
Cl_2 (aq)			
Br_2 (aq)			
I_2 (aq)/I⁻(aq)			

3. Formation of elemental halogens

	Reagents	Initial Observations (1a—3a)	Observations on Adding Hexane (1b—3b)	Conclusions
1	1 mL dil. H_2SO_4, 5 drops NaCl, 5 drops $NaClO_3$			
2	1 mL dil. H_2SO_4, 5 drops NaBr, 5 drops $NaBrO_3$			
3	1 mL dil. H_2SO_4, 5 drops NaI, 5 drops $NaIO_3$			

4. Formation of higher oxidation states of halogens

	Upon mixing of $NaIO_3$, NaI and HCl	Upon addition of hexane
phenomena		

5. The role of protons in the oxidation of iodide by iodate

	No. of drops before indicator changed colour
Test	
Blank	

6. The relative stabilities of the halogens in their −1 and 0 oxidation states

Tube	Reagents	Initial Observations (a)	Observations on adding hexane (b)	Conclusions
1				
2				
3				
4				
5				
6				

QUESTIONS

1. The properties of lithium, sodium and potassium salts

(1) What are the precipitates, if any, formed in this test ②? How does the

formation of precipitates depend upon hydration enthalpies and ionic radii?

(2) What are the precipitates, if any, formed in this test ③? How does the formation of precipitates depend upon hydration enthalpies and ionic radii? What effect can occur when a small cation interacts with a large anion?

(3) What are the precipitates, if any, formed in this test ④? How does the formation of precipitates depend upon hydration enthalpies and ionic radii? What is the important difference in this context between the ClO_4^- and F^- ions?

2. Colour of elemental halogens

(1) Look at the colours of the layers. Do you think halogens (X_2) are more soluble in organic or polar solvents? Explain your answer.

(2) If halogens are not very soluble in polar solvents—why do you think they produce such intense colours in water?

3. Formation of elemental halogens

(1) What chemical is produced initially in these tests?

(2) What happens to this chemical when you add hexane to the test tube?

(3) Write general half equations for the oxidation and reduction processes using X to represent the halogen. Write a balanced general overall equation for the formation of a halogen X_2.

4. Formation of higher oxidation states of halogens

(1) Write a balanced ionic equation for the oxidation of iodide by iodate in the presence of concentrated HCl.

(2) Explain briefly what is happening in test 4. How does the reaction differ from that observed in test 3?

5. The role of protons in the oxidation of iodide by iodate

(1) Balanced ionic equations for the reactions occurring in test 5

6. The relative stabilities of the halogens in their − 1 and 0 oxidation states

(1) Which of the above test tube tests lead to reaction?

(2) Write a balanced equation which summarises the findings from these tests.

(3) Explain what can be deduced concerning the trends in stability of the −1 and 0 oxidation states on descending Group 7A.

2.15 The Chemistry of Copper Group Elements and Quantitative Analysis by Titration

AIMS

1. To understand factors that affect the solubility of copper and silver salts.

2. To understand some basic co-ordination chemistry: the complexation of Ag^+ and Cu^{2+} in solution.

3. To learn the use of a redox titration as an analytical technique.

INTRODUCTION

Copper, silver, and gold are in group 11 of the periodic table; these three metals have one s-orbital electron on top of a filled d-electron shell and are characterized by high ductility, and electrical and thermal conductivity. Their electron configuration are similar, copper ($[Ar]3d^{10}4s^1$), silver($[Kr]4d^{10}5s^1$) and gold ($[Xe]4f^{14}5d^{10}6s^1$). Their physical and chemical properties are similar as well.

This distinctive electron configuration, with a single electron in the highest occupied s subshell over a filled d subshell, accounts for many of the singular properties of metals in group 11. The filled d-shells in these elements contribute little to interatomic interactions, which are dominated by the s-electrons through metallic bonds. That is, their single s electron is free and does not interact with the filled d subshell. Unlike metals with incomplete d-shells, metallic bonds in metals of group 11 are lacking a covalent character and are relatively weak. This observation explains the low hardness, high ductility, very high electrical and thermal conductivity of metals in group 11.

At the macroscopic scale, introduction of extended defects to the crystal lattice, such as grain boundaries, hinders flow of the material under applied stress, thereby increasing the metal's hardness. For this reason, copper is usually supplied in a fine-grained polycrystalline form, which has greater strength than monocrystalline forms.

Copper forms a rich variety of compounds, usually with oxidation states $+1$ and$+2$, which are often called cuprous and cupric, respectively. As with other elements, the simplest compounds of copper are binary compounds, i. e. those containing only two elements, the principal examples being oxides, sulfides, and halides.

Copper forms coordination complexes with ligands. In aqueous solution, copper (II) exists as $[Cu(H_2O)_6]^{2+}$. This complex exhibits the fastest water exchange rate (speed of water ligands attaching and detaching) for any transition metal aquo complex. Adding aqueous sodium hydroxide causes the precipitation of light blue solid copper(II) hydroxide. A simplified equation is:

$$Cu^{2+} + 2OH^- \Longrightarrow Cu(OH)_2 \tag{2-15-1}$$

Aqueous ammonia results in the same precipitate. Upon adding excess ammonia, the precipitate dissolves, forming tetraamminecopper(II):

$$Cu(H_2O)_4(OH)_2 + 4NH_3 \Longrightarrow [Cu(H_2O)_2(NH_3)_4]^{2+} + 2H_2O + 2OH^-$$
$$\tag{2-15-2}$$

Many other oxyanions form complexes; these include copper(II) acetate, copper (II) nitrate, and copper(II) carbonate. Copper(II) sulfate forms a blue crystalline pentahydrate, the most familiar copper compound in the laboratory. It is used in a fungicide called the Bordeaux mixture.

Silver is a soft, white, lustrous transition metal, it exhibits the highest electrical conductivity, thermal conductivity, and reflectivity of any metal. The metal is found in the Earth's crust in the pure, free elemental form ("native silver"), as an alloy with gold and other metals, and in minerals such as argentite and chlorargyrite. Most silver is produced as a byproduct of copper, gold, lead, and zinc refining.

Silver has rather low chemical affinities for oxygen, lower than copper, and it is therefore expected that silver oxides are thermally quite unstable. Soluble silver(I) salts precipitate dark-brown silver(I) oxide, Ag_2O, upon the addition of alkali. (The hydroxide AgOH exists only in solution; otherwise it spontaneously decomposes to the oxide.)

All four silver(I) halides are known. The fluoride, chloride, and bromide have the sodium chloride structure, but the iodide has three known stable forms at different temperatures; that at room temperature is the cubic zinc blende structure. They can all be obtained by the direct reaction of their respective elements. As the halogen group is descended, the silver halide gains more and more covalent character, solubility decreases, and the color changes from the white chloride to the yellow iodide as the energy required for ligand-metal charge transfer ($X^- Ag^+ \rightarrow XAg$) decreases. Except silver fluoride, the other three silver halides are highly insoluble in aqueous solutions and are very commonly used in gravimetric analytical methods. All four are photosensitive (though the monofluoride is so only to ultraviolet light), especially the bromide and iodide which photodecompose to silver metal, and thus were used in traditional photography.

Silver complexes tend to be similar to those of its lighter homologue copper. Like the valence isoelectronic copper(II) complexes, silver(II) complexes are usually square planar and paramagnetic, which is increased by the greater field splitting for 4d electrons than for 3d electrons. However, the most important oxidation state for silver in complexes is +1. Ag^+ forms salts with most anions, but it is reluctant to coordinate to oxygen and thus most of these salts are insoluble in water; the exceptions are the nitrate, perchlorate, and fluoride. The tetracoordinate tetrahedral aqueous ion $[Ag(H_2O)_4]^+$ is known, but the characteristic geometry for the Ag^+ cation is 2-coordinate linear. For example, silver chloride dissolves readily in excess aqueous ammonia to form $[Ag(NH_3)_2]^+$; silver salts are dissolved in photography due to the formation of the thiosulfate complex $[Ag(S_2O_3)_2]^{3-}$; and cyanide extraction for silver works by the formation of the complex $[Ag(CN)_2]^-$.

In this experiment you are going to test the properties of copper and silver salts firstly, pay attention to the characteristics and phenomena of reactions. Second, determine the amount of copper in an unknown copper salt by titration. You will then use the percentage copper to predict the nature of the salt. The copper(II) salt is dissolved and allowed to react with potassium iodide in acidic solution. The iodine released is titrated against standardised sodium thiosulphate solution.

REAGENTS AND APPARATUS

- Analytical balance, 100 mL beakers, 50 mL burettes, 250 mL conical beakers, 10 mL graduated cylinder, tubes, droppers.
- $0.1\ mol \cdot L^{-1}$ of silver nitrate ($AgNO_3$) solution, $0.1\ mol \cdot L^{-1}$ of copper sulphate($CuSO_4$) solution, copper salt, $0.1\ mol \cdot L^{-1}$ sodium chloride ($NaCl$) solution, $0.1\ mol \cdot L^{-1}$ sodium bromide ($NaBr$) solution, $0.1\ mol \cdot L^{-1}$ sodium iodide (NaI) solution, $0.1\ mol \cdot L^{-1}$ sodium chlorate ($NaClO_3$) solution, $0.1\ mol \cdot L^{-1}$ sodium bromate ($NaBrO_3$) solution, $0.1\ mol \cdot L^{-1}$ sodium iodate ($NaIO_3$) solution, $0.1\ mol \cdot L^{-1}$ ammonia water ($NH_3 \cdot H_2O$), potassium iodide (KI), $0.1\ mol \cdot L^{-1}$ sodium thiosulphate ($Na_2S_2O_3$) solution, 1% starch solution, hexane, distilled water.

PROCEDURES

1. The interaction of Ag(I) with various anions

Number three test tubes 1 to 3. To each test tube add 1 mL $0.1\ mol \cdot L^{-1}$ of $AgNO_3$ solution. To each test tube add 5 to 10 drops of a solution of $0.1\ mol \cdot L^{-1}$ $NaCl$, $NaBr$, and NaI, seperately. Decant the liquid from each test tube and add 2 mL $0.1\ mol \cdot L^{-1}$ $NH_3 \cdot H_2O$.

2. Interaction of Cu(II) with various anions

To test tubes labelled 4 to 10 add 1 mL 0.1 mol $\cdot$ L^{-1} CuSO$_4$ solution. To test tube 4—9, add 1 mL of a solution of 0.1 mol $\cdot$ L^{-1} NaCl, NaBr, NaI, NaClO$_3$, NaBrO$_3$, NaIO$_3$, seperately. Then add 2 mL of 0.1 mol $\cdot$ L^{-1} NH$_3$ $\cdot$ H$_2$O solution to each tube. Add 1 mL of a solution of 0.1 mol $\cdot$ L^{-1} NaI to test tube 10, followed by 1 mL hexane; mix well and note the colour of both layers.

3. Determination of Copper

Into each of three 250 mL conical flasks weigh accurately about 0.4 g of the given copper compound (four significant digits after decimal point), add 100 mL distilled water to dissolve the sample. If the sample fails to dissolve, add glacial acetic acid dropwise until the sample just dissolves. Add about 1 g of potassium iodide (four significant digits after decimal point) to the solution and titrate the liberated iodine with the 0.1 000 mol $\cdot$ L^{-1} standardised sodium thiosulphate solution until a pale straw colour is obtained. Add 1 mL 1% starch solution and continue the titration until the blue colour is discharged. The endpoint should be stable for 15 to 20 seconds but will revert to the blue colour on standing, due to oxidation by the air. Parallel repeated determination three times.

DATA TREATMENT

1. The interaction of Ag(I) with various anions

Tube	Addition of NaX	Observations	Observation after adding NH$_3$ $\cdot$ H$_2$O
1			
2			
3			

2. Interaction of Cu(II) with various anions

Tube	Addition of NaX	Observations	Observation after adding NH$_3$ $\cdot$ H$_2$O/hexane
4	NaCl		
Eq'n			
5	NaBr		
Eq'n			

(**Continued**)

Tube	Addition of NaX	Observations	Observation after adding $NH_3 \cdot H_2O$/hexane
6	NaI		
Eq'n			
7	NaClO₃		
Eq'n			
8	NaBrO₃		
Eq'n			
9	NaIO₃		
Eq'n			
10	NaI		
Eq'n			

3. Determination of Copper

		1	2	3
CuX weighting	m(CuX)/g			
KI weighting	m(KI)/g			
Na₂S₂O₃ titrant	before titration/mL			
	after titration/mL			
	net volume/mL			
% Cu				
average% Cu				
the average relative deviation				
Formula of copper salt analysed				

QUESTIONS

1. Using X^- to represent Cl^-, Br^- and I^-, write balanced equations to explain both the equilibria you observe in the stop of "The interaction of Ag(I)". (i. e. Ag^+/X^- and AgX/NH_3).

2. What is the shape and stoichiometry of the complex silver ion formed in

solution?

3. Write a balanced ionic equation for the reduction of Cu(II) to Cu(I) by iodide (I^-) and the subsequent formation of copper(I) iodide and iodine. Write a balanced equation for the reduction of the released iodine by sodium thiosulphate solution $(Na_2S_2O_3)$ with the formation of the tetrathionate ion $(S_4O_6^{2-})$.

4. What are the factors that affect the solubility of copper and silver salts?

2.16 The Chemistry of Nitrogen and Phosphorous

AIMS

1. To master the main properties of nitrogen compounds.
2. To know the acidity and basicity, and solubility of phosphate.

INTRODUCTION

Nitrogen is a chemical element with symbol N and atomic number 7. A nitrogen atom has seven electrons. In the ground state, they are arranged in the electron configuration of $1s^2 2s^2 2p_x^1 2p_y^{1\,2} p_z^1$. It therefore has five valence electrons in the 2s and 2p orbitals, three of which (the p-electrons) are unpaired. It has one of the highest electronegativities among the elements. Because the high electronegativity makes it difficult for a small nitrogen atom to be a central atom in an electron-rich three-center four-electron bond since it would tend to attract the electrons strongly to itself.

Ammonia (NH_3) is the most important compound of nitrogen and is prepared in larger amounts than any other compound, because it contributes significantly to the nutritional needs of terrestrial organisms by serving as a precursor to food and fertilisers. As a liquid, it is a very good solvent with a high heat of vaporisation, a low viscosity and electrical conductivity, and high dielectric constant, and is less dense than water. The ammonium cation is a positively charged polyatomic ion with the chemical formula NH_4^+. It is formed by the protonation of ammonia (NH_3). Ammonium is also a general name for positively charged or protonated substituted amines and quaternary ammonium cations $(NR_4^+ 4)$, where one or more hydrogen atoms are replaced by organic groups (indicated by R). Ammonium cation is found in a variety of salts such as ammonium carbonate, ammonium chloride, and ammonium nitrate. Most simple ammonium salts are very soluble in water. The ammonium salts of nitrate and especially perchlorate are highly explosive, in these cases ammonium is the reducing agent.

Many nitrogen oxoacids are known, though most of them are unstable as pure compounds and are known only as aqueous solution or as salts. Nitrous acid (HNO_2) is not known as a pure compound, but is a common component in gaseous equilibria and is an important aqueous reagent: its aqueous solutions may be made from acidifying cool aqueous nitrite (NO_2^-, bent) solutions. It is a weak acid with pK_a of 3. 35 at 18 ℃. They may be titrimetrically analysed by their oxidation to nitrate by permanganate. They are readily reduced to nitrous oxide and nitric oxide by sulfur dioxide, to hyponitrous acid with tin(II), and to ammonia with hydrogen sulfide. Sodium nitrite is mildly toxic in concentrations above 100 mg • kg^{-1}, but small amounts are often used to cure meat and as a preservative to avoid bacterial spoilage.

Nitric acid (HNO_3) is by far the most important and the most stable of the nitrogen oxoacids. It is one of the three most used acids (the other two being sulfuric acid and hydrochloric acid). It is made by catalytic oxidation of ammonia to nitric oxide, which is oxidised to nitrogen dioxide, and then dissolved in water to give concentrated nitric acid, most of which is used for nitrate production for fertilisers and explosives, among other uses. It is a strong acid and concentrated solutions are strong oxidising agents, though gold, platinum, rhodium, and iridium are immune to attack. A 3 : 1 mixture of concentrated hydrochloric acid and nitric acid is still stronger and successfully dissolves gold and platinum. The thermal stabilities of nitrates (involving the trigonal planar NO_3^- anion) depends on the basicity of the metal, and so do the products of decomposition (thermolysis). Nitrate is also a common ligand with many modes of coordination.

Phosphorus is a chemical element with symbol P and atomic number 15. As an element, phosphorus exists in two major forms—white phosphorus and red phosphorus—but because it is highly reactive, phosphorus is never found as a free element on earth. It is thesecond element in pnictogen column. Phosphorus is essential for life. Phosphates (compounds containing the phosphate ion, PO_4^{3-}) are a component of DNA, RNA, ATP, and the phospholipids, which form all cell membranes. The bulk of all phosphorus production is in concentrated phosphoric acids for agriculture fertilisers. Other uses are including antirust converter, food additive, dental and orthopedic etchant, electrolyte, soldering flux, dispersing agent, industrial etchant, and so on.

The most prevalent compounds of phosphorus are derivatives of phosphate (PO_4^{3-}), a tetrahedral anion. It is a non-toxic acid, which, when pure, is a solid at room temperature and pressure. Being triprotic, phosphoric acid converts stepwise to three conjugate bases:

$$H_3PO_4 + H_2O • H_3O^+ + H_2PO_4^- \qquad K_{a1} = 7. 25 \times 10^{3-} \qquad (2\text{-}16\text{-}1)$$

$$H_2PO_4^- + H_2O \cdot H_3O^+ + HPO_4^{2-} \qquad K_{a2} = 6.31 \times 10^{8-} \qquad (2\text{-}16\text{-}2)$$

$$HPO_4^{2-} + H_2O \cdot H_3O^+ + PO_4^{3-} \qquad K_{a3} = 3.98 \times 10^{13-} \qquad (2\text{-}16\text{-}3)$$

Phosphate exhibits the tendency to form chains and rings with P-O-P bonds. Many polyphosphates are known, including ATP. Polyphosphates arise by dehydration of hydrogen phosphates such as HPO_4^{2-} and $H_2PO_4^-$.

Phosphate is the conjugate base of phosphoric acid. A phosphate salt forms when a positively charged ion attaches to the negatively charged oxygen atoms of the ion, formingan ionic compound. The sodium, potassium, rubidium, caesium, and ammonium phosphates are all water-soluble. Most other phosphates are only slightly soluble or are insoluble in water. As a rule, the hydrogen and dihydrogen phosphates are slightly more soluble than the corresponding phosphates. The pyrophosphates are mostly water-soluble.

Pyrophosphoric acid, is colorless, odorless, hygroscopic and is soluble in water, diethyl ether, and ethyl alcohol. It is a medium strong inorganic acid. Anions, salts, andesters of pyrophosphoric acid are called pyrophosphates. It is best prepared by ion exchange from sodium pyrophosphate or by reacting hydrogen sulfide with lead pyrophosphate. When phosphoric acid is dehydrated, pyrophosphoric acid is produced as one of the products. In aqueous solution pyrophosphoric acid, like all polyphosphoric acids, hydrolyses and eventually an equilibrium is established between phosphoric acid, pyrophosphoric acid, and polyphosphoric acids. When highly diluted an aqueous solution of pyrophosphoric acid contains only phosphoric acid.

In this experiment you are going to test the properties of some ammonium salt, nitrate, phosphate, and pay attention to the characteristics and phenomena of reactions.

REAGENTS AND APPARATUS

- Analytical balance, heating mantle, iron stand, iron jaws, water bath, 100 mL beakers, 10 mL graduated cylinder, watch glass, tubes, droppers, pH test paper.
- Ammonium chloride (NH_4Cl), ammonium sulfate (($NH_4)_2SO_4$), ammonium dichromate (($NH_4)_2Cr_2O_7$), sodium nitrate ($NaNO_3$), copper nitrate ($Cu(NO_3)_2$), silver nitrate ($AgNO_3$), sulfur powder (S), zinc chip (Zn), splint, 0.1 mol $\cdot$ L^{-1} sodium nitrite ($NaNO_2$) solution, 3 mol $\cdot$ L^{-1} sulfuric acid (H_2SO_4) solution, 0.1 mol $\cdot$ L^{-1} potassium iodide (KI) solution, 0.1 mol $\cdot$ L^{-1} potassium permanganate ($KMnO_4$) solution, 6 mol $\cdot$ L^{-1} and 0.5 mol $\cdot$ L^{-1} nitric acid (HNO_3) solution, 0.1 mol $\cdot$ L^{-1} barium chloride

(BaCl$_2$), 40% sodium hydroxide (NaOH) solution, 0. 1 mol • L^{-1} sodium phosphate (Na$_3$PO$_4$) solution, 0. 1 mol • L^{-1} sodium hydrogenphosphate (Na$_2$HPO$_4$) solution, 0. 1 mol • L^{-1} sodium dihydrogen phosphate (NaH$_2$PO$_4$) solutions, 0. 1 mol • L^{-1} silver nitrate (AgNO$_3$) solution, 0. 1 mol • L^{-1}calcium chloride (CaCl$_2$) solution, 2 mol • L^{-1} ammonia water (NH$_3$ • H$_2$O), 2 mol • L^{-1} hydrochloric acid (HCl) solution, 0. 2 mol • L^{-1} copper sulphate (CuSO$_4$) solution, 0. 1 mol • L^{-1} sodium pyrophosphate (Na$_4$P$_2$O$_7$) solution, 1% starch solution, distilled water.

PROCEDURES

1. Thermal decomposition of ammonium salts

Add 1 g ammonium chloride in a dry test tube. Fix the test tube vertically with an iron jaw. Heat the test tube with a heating mantle. Then place a wet pH test paper across the tube mouth to test the escaped gas, and observe the color change of the test paper. Continue heating, and pay attention to the color change of the pH test paper. At the same time, observe the phenomenon of the upper part of the tube wall.

Repeat the above experiments with ammonium sulfate and ammonium dichromate instead of ammonium chloride, respectively. Observe and compare the phenomena during the decomposition of these ammonium salts.

2. Nitrous acid and nitrite

2. 1   Formation and decomposition of nitrite

Place 1 mL 0. 1 mol • L^{-1} sodium nitrite solution in a test tube. The test tube is then cooled in ice water bath. Slowly add 1 mL 3 mol • L^{-1} sulfuric acid solution. Observe the reaction and the product's color. Take the test tube out of the ice water and put it on the test tube rack. Wait for a moment, and observe the phenomena in the test tube.

2. 2   Oxidation and reduction of nitrite

Add 1 mL 0. 1 mol • L^{-1} potassium iodide solution in a test tube, and acidify it by adding 2 drops of 3 mol • L^{-1} sulfuric acid solution. Drop 1 mL 0. 1 mol • L^{-1} sodium nitrite solution. And then add 1 mL 1% starch solution. Watch the phenomena, and write the reaction equation.

Repeat this redox reaction experiment, using 0. 1 mol • L^{-1} potassium permanganate solution instead of potassium iodide solution. Watch the phenomena, and write the reaction equation.

3. Nitric acid and nitrate

3.1 Oxidation of nitric acid

Place sulfur powder, as large as soybean, in a test tube, and then add 1 mL $6 \text{ mol} \cdot \text{L}^{-1}$ concentrated nitric acid. Be careful. Put the test tube in the water bath at 80 ℃. Observe the reaction, and pay attention to the gas color. After reaction, cool to room temperature. Add 1 mL $0.1 \text{ mol} \cdot \text{L}^{-1}$ barium chloride in the test tube to check the products.

Put a piece of zinc chip in each of the two test tubes. One tube is adding 1 mL $6 \text{ mol} \cdot \text{L}^{-1}$ concentrated nitric acid, and another is adding 1 mL $0.5 \text{ mol} \cdot \text{L}^{-1}$ dilute nitric acid. Observe and compare the reaction rates and products. Add two drops of the products from the tube containing dilute nitric acid onto a watch glass. Then add a drop of 40% concentrated sodium hydroxide, and quickly cover another watch glass upside down, which is attached with a wet red litmus test paper. Put this set of watch glasses in a heating mantle. Observe if the red litmus test paper turns blue.

3.2 Thermal decomposition of nitrates

Put a small amount of sodium nitrate into a dried test tube. Heat the test tube with the heating mantle, and observe the reaction and the color of the product. Place a glowing splint at the tube mouth, watch whether it can relight. Repeat this experiment with copper nitrate and silver nitrate instead of sodium nitrate, respectively.

4. Properties of phosphates

4.1 Acidity and basicity

Determine the pH of $0.1 \text{ mol} \cdot \text{L}^{-1}$ Na_3PO_4, Na_2HPO_4 and NaH_2PO_4 solutions by pH test paper.

Add 0.5 mL $0.1 \text{ mol} \cdot \text{L}^{-1}$ Na_3PO_4, Na_2HPO_4 and NaH_2PO_4 solutions into three test tubes, respectively. Add 1 mL $0.1 \text{ mol} \cdot \text{L}^{-1}$ $AgNO_3$ solution to each test tube. Then, test the pH of the tubes with pH test paper after reactions. Watch the phenomena, and write the reaction equation.

4.2 Solubility

Add 0.5 mL $0.1 \text{ mol} \cdot \text{L}^{-1}$ Na_3PO_4, Na_2HPO_4 and NaH_2PO_4 solutions into three test tubes, respectively. Drop 1 mL $0.1 \text{ mol} \cdot \text{L}^{-1}$ calcium chloride solution. Observe the phenomena. Test the pH of the tubes with pH test paper after reactions. Drop a few drops of $2 \text{ mol} \cdot \text{L}^{-1}$ ammonia water. Observe the phenomena. Then drop a few drops of $2 \text{ mol} \cdot \text{L}^{-1}$ hydrochloric acid. Observe the phenomena.

4.3 Coordination

Add 0.5 mL $0.2 \text{ mol} \cdot \text{L}^{-1}$ $CuSO_4$ solution into a test tube, and then add 0.5 mL $0.1 \text{ mol} \cdot \text{L}^{-1}$ sodium pyrophosphate solution drop by drop. Observe the formation of precipitation. Continue to drop 1. 5 mL 0. 1 mol $\cdot$ L^{-1} sodium pyrophosphate solution, Observe whether the precipitation is dissolved. Write the corresponding reaction equation.

DATA TREATMENT

1. Thermal decomposition of ammonium salts. Observation of experimental phenomena.

	pH test paper	Top of tube	Tube bottom
NH$_4$Cl			
Eq'n			
(NH$_4$)$_2$SO$_4$			
Eq'n			
(NH$_4$)$_2$Cr$_2$O$_7$			
Eq'n			

2. Nitrous acid and nitrite

2. 1   Formation and decomposition of nitrite. Observation of experimental phenomena.

	In ice-water bath	Out of ice-water bath
NaNO$_2$ + H$_2$SO$_4$		
Eq'n		

2. 2 Oxidation and reduction of nitrite. Observation of experimental phenomena.

	Adding NaNO$_2$	Adding starch solution
KI (H$^+$)		
Eq'n		
KMnO$_4$ (H$^+$)		
Eq'n		

3. Nitric acid and nitrate

3. 1 Oxidation of nitric acid. Observation of experimental phenomena.

	In water bath	Adding BaCl$_2$
S + HNO$_3$		
Eq'n		

	Zn	40% NaOH
Conc. HNO_3		
Eq'n		
Dilute HNO_3		
Eq'n		

3.2 Thermal decomposition of nitrates. Observation of experimental phenomena.

	Δ	Glowing splint
$NaNO_3$		
Eq'n		
$Cu(NO_3)_2$		
Eq'n		
$AgNO_3$		
Eq'n		

4. Properties of phosphates

4.1 Acidity and basicity. Observation of experimental phenomena.

	pH test paper	Adding $AgNO_3$	pH test paper
Na_3PO_4			
Eq'n			
Na_2HPO_4			
Eq'n			
NaH_2PO_4			
Eq'n			

4.2 Solubility. Observation of experimental phenomena.

	Adding $CaCl_2$	pH test paper	Adding $NH_3 \cdot H_2O$	Adding HCl
Na_3PO_4				
Eq'n				
Na_2HPO_4				
Eq'n				
NaH_2PO_4				
Eq'n				

4.3   Coordination. Observation of experimental phenomena.

	Adding $Na_4P_2O_7$	Continue to add $Na_4P_2O_7$
$CuSO_4$		
Eq'n		

QUESTIONS

1. How to distinguish nitrate from nitrite.

2. Which acid among hydrochloric acid, sulphuric acid and nitric acid, is the best choice to dissolve silver phosphate precipitation? Why?

3. How to identify sodium phosphate, sodium hydrogen phosphate and sodium dihydrogen phosphate? List specific methods.

2.17   Identification of Unknown Compounds

AIMS

1. To learn how to identify common substances with the basic properties of the elements and compounds.

2. To review important cationic andanionic reactions.

INTRODUCTION

This experiment requires you to put your general chemical knowledge into practice. All the information you need is in any textbook of qualitative analysis. Some points should be obvious, such as that a white solid cannot be HCl, and a black solid cannot be NaCl or CaO. Similarly, a white solid which cannot be dissolved in water is not NaCl. A white solid which reacts exothermically with water, dissolves only slightly, and which is soluble in dilute HCl, is very likely to be CaO.

Solubility is very important as a means of identification, since the number of readily soluble salts is rather small. Thus, if a material is soluble; this is useful information already. If you can find some precipitation reactions for the solution, you have a lot more information.

While there are no firm rules about solubility, since this depends on a delicate balance between the free energies of formation of the solid and solution, some rough guides can be found, which are some help in memorising the basic knowledge expected of any chemist. Generally the very insoluble combinations are those where a large lattice energy is expected. Virtually all Group I salts are soluble, and so are many halides, though Ag halides are an important exception. If the solution of x gives a white precipitate with $AgNO_3$, then it is probably a chloride.

Many $2+/2-$ combinations, e. g. $CaSO_4$, $BaSO_4$ have low solubilities (exception $MgSO_4$), and $2+/3-$ and $3+/2-$ combinations are normally very insoluble, particularly if the ions are small. An apparent exception is the very soluble aluminium sulphate, used for water treatment. This is due to the high enthalpy of hydration of the metal cation and the same effect is found for Mg^{2+}.

Many solutions will react with others to give precipitates, so for example if you suspect that two of your compounds are Na_2SO_4 and $CaCl_2$, then if both dissolve, and mixing the two solutions gives a white precipitate $(2+/2-)$, you have evidence to support this idea. However, if the precipitate is yellow, then you need to think again.

Once you have some idea of what's going on, then flame tests, which characterize some metal ions by the specific colours of the metal atom emission spectra, can be useful (all atoms have these spectra, but only a few have strong lines in the tiny region of the spectrum which we can see). The significant ones are Na, which gives the same intense yellow as a sodium street lamp. K gives lilac, Cu a blue, and Ba a green flame. Ca gives a rather weak red flame ("brick-red" according to the books), Sr a deep crimson. Next time you watch fireworks, you should be able to tell what went into them.

The Principles of Sorting: 1) Observation of state and colour; 2) Water solubility of each of the solids-high, slight or none (do you expect the black solids to dissolve?); 3) Initial reactions of liquids with each other or with solutions from; 4) Once you have some idea from these you can try flame tests.

Example: Suppose you are given H_2O, dil HCl, $AgNO_3$ solution, (all colourless liquids), and the solids: $NiCl_2$, K_2CO_3, $CaSO_4$, and CuO. The last of these is black and $NiCl_2$ is green. You can check that it is $NiCl_2$ by dissolving some in water and adding the solution to each of the three liquids in turn-only one of them will react-acid has no effect but silver nitrate will give a precipitate.

Now that you have identified $AgNO_3$, you have a very useful reagent; in particular you have a test for Cl^- and so you can distinguish HCl from H_2O.

Now that you have HCl, you have an acid. This can be used to find carbonates and will also dissolve many metal oxides. In this case, you should already have a good idea which is the potassium carbonate, since it is the only water-soluble white solid. As a final confirmation, the white, water-soluble material, which fizzes with acid, also gives a lilac flame test, and the insoluble one, gives a reddish flame.

An alternative, equally valid, but trickier procedure to get this far, would have been to try all the possible combinations of the liquids, which tells you which pair is HCl/AgNO$_3$, and then try each of these with the two white solids.

In this experiment, you should be able to draft your strategy in your laboratory notebook before embarking on the tests. A flow diagram is a good way to represent this. Record all your observations, deductions, and final conclusions in your laboratory notebook.

REAGENTS AND APPARATUS

- Test tubes, droppers, reagent spoons.
- Solid: Aluminium oxide (Al_2O_3), aluminium sulphate ($Al_2(SO_4)_3$), ammonium sulphate (($NH_4)_2SO_4$), barium chloride ($BaCl_2$), calcium nitrate ($Ca(NO_3)_2$), calcium oxide (CaO), copper(II) oxide (CuO), copper(II) sulphate ($CuSO_4$), magnesium sulphate ($MgSO_4$), manganese(IV) oxide (MnO_2), potassium iodide (KI), zinc carbonate ($ZnCO_3$), sodium carbonate (Na_2CO_3), sodium dithionite ($Na_2S_2O_4$), sodium thiosulphate ($Na_2S_2O_3$).
  Liquid: 2 mol • L^{-1} ammonium hydroxide (NH_4OH), 2 mol • L^{-1} sulphuric acid (H_2SO_4), 0.1 mol • L^{-1} silver nitrate ($AgNO_3$) solution, distilled water.

PROCEDURES

1. Observation of state, color, and odor.
2. Water solubility of each of the solids. Pay attention to if there is some precipitation or bubbles during the dissolution process.
3. Reactions of samples with each other. Pay attention to if there is some precipitation orbubbles, or color changes during the dissolution process.
4. Once you have some ideas from these you can try the flame tests.

DATA TREATMENT

1. Produce a neat and legible version of flow chart.
2. Identification of samples.

Sample number	Code No.	Identification	Comments

QUESTIONS

1. Write out the reaction equations used to identify samples.

2. Which samples are particularly difficult to identify during the experiment? How to solve them finally?

2.18   Preparation and Quantitative Analysis of Sodium Thiosulfate

1. To master the reaction principle and control of reaction conditions of sodium thiosulfate synthesis.
2. To master the method of quantitative analysis of products.
3. Basic operational skills of related comprehensive experiments.

INTRODUCTION

Sodium thiosulfate, which is also referred to as sodium sulphate, is a chemical compound which has the formula $Na_2S_2O_3$. It is typically found in its pentahydrate form which is either white in color, or colorless altogether. This pentahydrate of sodium thiosulfate is described by the following chemical formula: $Na_2S_2O_3 \cdot 5H_2O$.

In its solid form, it is a crystalline solid which has a tendency to readily lose water. Sodium thiosulfate is readily soluble in water and is also referred to as sodium hyposulfite. The structure of the $Na_2S_2O_3$ molecule is illustrated below.

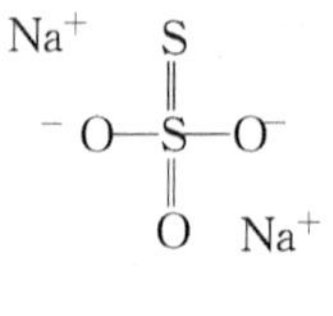

$$Na^+ \quad \overset{\displaystyle S}{\underset{\displaystyle O \quad Na^+}{\overset{\|}{\underset{\|}{^-O-S-O^-}}}}$$

Figure 2 – 24   The structure of $Na_2S_2O_3$

It can be noted that the shape of the thiosulfate ion is tetrahedral in the solid state of sodium thiosulfate. The distance between the two sulfur atoms in the thiosulfate ion is comparable to the distance between two sigma bonded sulfur atoms. This implies that the sulfur which is not bonded to any oxygens holds a negative charge.

The physical and chemical properties of $Na_2S_2O_3$ are listed below:

Physical Properties: It has a white, crystalline appearance as a solid and is odorless. The crystal structure of $Na_2S_2O_3$ crystals is monoclinic. In its anhydrous form, $Na_2S_2O_3$ has a molar mass of 158.11 g $\cdot$ mol^{-1}. The more commonly available pentahydrate from, $Na_2S_2O_3 \cdot 5H_2O$ has a molar mass of 248.18 g $\cdot$ mol^{-1}. The density of sodium thiosulfate corresponds to 1. 667 g • mL^{-1}. The pentahydrate of this salt has a melting point of 321. 4 K and a boiling point of 373 K. The solubility of sodium thiosulfate in water is 70. 1g/100 mL at 20 ℃ and 231g/100 mL at 100 ℃.

Chemical Properties: The sodium thiosulfate salt is neutral in charge. However, it dissociates in water and some other polar solvents to yield Na^+ and $S_2O_3^{2-}$. Despite being stable at standard conditions, the sodium thiosulfate salt decomposes at high temperatures to yield sodium sulfate along with sodium polysulfide. The chemical equation for the reaction described above is given by:

$$4Na_2S_2O_3 =\!=\!= 3Na_2SO_4 + Na_2S_5 \tag{2-18-1}$$

When exposed to dilute acids such as dilute hydrochloric acid, the sodium thiosulfate salt undergoes a decomposition reaction to yield sulfur along with sulfur dioxide.

$$Na_2S_2O_3 + 2HCl =\!=\!= 2NaCl + SO_2 + H_2O + S \tag{2-18-2}$$

The alkylation of $Na_2S_2O_3$ yields S-alkyl thiosulfates. These compounds are commonly referred to as Bunte salts.

Production of sodium thiosulfate

The laboratory preparation of the sodium thiosulfate salt involves the heating of aqueous sodium sulfite solutions along with sulfur. The production of $Na_2S_2O_3$ can also be accomplished by the boiling of aqueous NaOH (sodium hydroxide) with sulfur.

The reaction for the method described above is given by:

$$6NaOH + 4S =\!=\!= Na_2S_2O_3 + 2Na_2S + H_2O \tag{2-18-3}$$

Industrially, sodium thiosulfate is prepared from the liquid waste generated from the manufacture of sulfur dye.

It can be noted that upon heating with Al^{3+} containing samples, sodium thiosulfate gives a white colored precipitate. This is due to the reaction between the aluminum cation and the thiosulfate anion which forms aluminum hydroxide along with sulfur and sulfur dioxide.

Some important uses of sodium thiosulfate

$Na_2S_2O_3$ is a very important chemical compound in the medical treatment of cyanide poisoning cases. Sodium thiosulfate is also used medically to treat dermatophytosis (ringworm) and tinea versicolor. The side effects of chemotherapy and hemodialysis (purification of blood) are treated with $Na_2S_2O_3$. It is a very important compound in analytical chemistry since it stoichiometrically reacts with iodine to reduce it to the iodide ion while it is oxidized to the $S_4O_6^{2-}$ (tetrathionate) ion. Sodium thiosulfate salts are used as photographic fixers due to the ability of the thiosulfate ions to react with silver halides, which make up photographic emulsions.

Ammonium thiosulfate and sodium thiosulfate make up lixiviants (liquid mediums used in hydrometallurgy) which are used in the extraction of gold from its ores. In order to reduce the chlorine levels in water, $Na_2S_2O_3$ is used in the dechlorination process. Thus, it is evident that sodium thiosulfate is a very important chemical compound in the lives of human beings.

Quantitative analysis of sodium thiosulfate

Iodine can be used as an oxidizing agent in many oxidation-reduction titrations and iodide can be used as a reducing agent in other oxidation-reduction titrations:

$$I_2 + 2e^- = 2I^- \tag{2-18-4}$$

If a standard iodine solution is used as a titrant for an oxidizable analyte, the technique is iodimetry. If an excess of iodide is used to quantitatively reduce a chemical species while simultaneously forming iodine, and if the iodine is subsequently titrated with thiosulfate, the technique is iodometry. Iodometry is an example of an indirect determination since a product of a preliminary reaction is titrated.

The use of iodine as a titrant suffers from two major disadvantages. First, iodine is not particularly soluble in water, and second, iodine is somewhat volatile. Consequently, there is an escape of significant amounts of dissolved iodine from the solution. Both of these disadvantages are overcome by adding iodide (I^-) to iodine (I_2) solutions. In the presence of iodide, iodine reacts to form triiodide (I^-) which is highly soluble and not volatile.

$$I_2 + I^- = I_3^- \tag{2-18-5}$$

The major chemical species present in these solutions is triiodide. The reduction of triiodide to iodide is analogous to the reduction of iodine.

$$I_3^- + 2e^- = 3I^- \tag{2-18-6}$$

Triiodide reacts with thiosulfate to yield iodide and tetrathionate.

$$2S_2O_3^{2-} + I_3^- = S_4O_6^{2-} + 3I^- \tag{2-18-7}$$

Dilute triiodide solutions are yellow, more concentrated solutions are brown, and even more concentrated solutionsare violet. Iodide solutions are colorless. If all of the other solution components are colorless, it is possible to detect the endpoint of titrations involving triiodide without the use of an indicator. Endpoint detection is considerably easier, however, with an indicator. The indicator that is usually chosen for titrations involving iodine (triiodide) is starch. Starch forms a dark blue complex with iodine. The endpoint in iodimetry corresponds to a sudden color change to blue. Likewise the endpoint in iodometry corresponds to a sudden loss of blue color due to the complex. Potato starch, rather than corn starch, is preferred for making the indicator solution since the color change due to the starch complex at the endpoint is sharper. In iodometry the starch is added only after the color due to triiodide has begun to fade, i. e., near the endpoint, because starch can be destroyed in the presence of excess triiodide.

Potassium iodate is a primary standard. Iodate (IO_3^-) reacts with an excess of iodide in acid solution to yield triiodide, which is subsequently titrated with the thiosulfate solution.

$$IO_3^- + 8I^- + 6H^+ = 3I_3^- + 3H_2O \qquad (2\text{-}18\text{-}8)$$

The standardization is an example of iodometry.

REAGENTS AND APPARATUS

- Analytical balance, heating mantle with stirrer, stir bars, oven, water pump, 100 mL beakers, 50 mL burettes, 250 mL volumetric flasks, 250 mL conical beakers, 10 mL and 50 mL graduated cylinder, brinell funnel, 500 mL filter bottle, evaporating dishes, surface dishes, glass rods
- Sulphur powder (S), sodium sulphite ($NaSO_3$), potassium iodate (KIO_3), potassium iodide (KI), 6 mol $\cdot$ L^{-1} hydrochloric acid (HCl), 2 g $\cdot$ L^{-1} starch solution, ethanol, distilled water

PROCEDURES

1. Preparation of sodium thiosulfate

2 g sulphur powder is weighed and placed in a 100 mL beaker. 1 mL ethanol is added to make the sulphur powder wet. 8.0 g anhydrous sodium sulphite (relative molecular weight 126) is then weighed and added to the beaker. Add 50 mL distilled water.

Heat until boiling under continuous stirring, and keep the boiling state for 30 minutes. During the heating process, pay attention to adding water in time, and the total volume should remain unchanged at 50mL. After the reaction is completed, 1—2 g activated carbon is added to the boiling solution. The activated carbon is used as decolorizer to adsorb excess sulfur powder. Continue boiling for about 10 mins under constant stirring.

Filter while the solution is hot, wash with a little distilled water three times, discard the solid, and retain the filter liquor. The filtrate is then put in an evaporation dish and heated to faint boiling. When the filtrate concentrates to the point where crystallization begins to show, stop the heating. After cooling to the room temperature, you will find a lot of white crystals in the evaporation dish.

Filter and wash with very little ethanol, retain the crystallization. The white crystallization is pentahydrate sodium thiosulfate ($Na_2S_2O_3 \cdot 5H_2O$), which is then put in a surface dish. Place the surface dish is the oven, and dry at 40 ℃ for 40— 60 mins.

Weigh and calculate the yield (theoretical yield is 15. 75 g).

2. Quantitative analysis of sodium thiosulfate

(1) Preparation of 0. 1 mol $\cdot$ L^{-1} $Na_2S_2O_3$ solution

Weight the prepared $Na_2S_2O_3$ $\cdot$ $5H_2O$ about 6. 2 g and record the mass to 4 decimal places, dissolve it in with distilled water in 100 mL beaker. Transfer the solution to a 250. 0 mL volumetric flask, dilute it with water to the calibration mark.

The burette is washed with water and distilled water three times, and rinsed with the prepared $Na_2S_2O_3$ solution, and then filled with $Na_2S_2O_3$ solution.

(2) Calibration of 0. 1 mol $\cdot$ L^{-1} $Na_2S_2O_3$ solution

Weight accurately 0. 1 g KIO_3 in a 250 mL conical beaker and record the mass to 4 decimal places. 30 mL distilled water is added to dissolve KIO_3 completely. Repeat three times.

Weight accurately 4. 15 g KI and record the mass to 4 decimal places. Dissolve it in a 100 mL beaker with 50 mL distilled water, and then transfer the solution to a 250. 0 mL volumetric flask. Dilute with water to the calibration mark.

Before titration, add 25 mL KI solution and 5 mL 6 mol $\cdot$ L^{-1} HCl solution to the KIO_3 conical beaker. Shake well. Titrate immediately with $Na_2S_2O_3$ solution until the color turns to light yellow.

Add 5 mL 2 g $\cdot$ L^{-1} starch solution. When starch is added, the color of the solution becomes red. If the quality of starch is poor, the color becomes light purple. Continue to titrate $Na_2S_2O_3$ solution until the color of solution is just colorless, that is the endpoint. Individually repeat the titration steps with the remaining two potassium iodate samples. To minimize error, generate the triiodide just before you titrate it with thiosulfate.

According to the weight of KIO_3 and the volume of $Na_2S_2O_3$ solution consumed, the concentration of $Na_2S_2O_3$ solution can be calculated.

DATA TREATMENT

1. Preparation of sodium thiosulfate

	reactant: Na_2SO_3	product: $Na_2S_2O_3$	yield
weight			

2. Quantitative analysis of sodium thiosulfate

		1	2	3
$Na_2S_2O_3$ weighting	$m(Na_2S_2O_3 \cdot 5H_2O)/g$			
KIO_3 weighting	$m(KIO_3)/g$			
$Na_2S_2O_3$ titrant	beforetitration/mL			
	after titration/mL			
	net volume/mL			
$c(Na_2S_2O_3)/mol \cdot L^{-1}$				
average $\bar{c}(Na_2S_2O_3)/mol \cdot L^{-1}$				
the average relative deviation				

QUESTIONS

1. Which reactant is excess, sulfur powder or sodium sulphite? Why?

2. In the final filtration for $Na_2S_2O_3$ preparation, why use ethanol to wash the crystallization?

3. What is the purpose of adding activated carbon?

4. Why should the drying temperature of oven be controlled at 40 ℃?

5. In the titration step, why should starchsolution be added when the color of the solution is light yellow?

2.19 Preparation and Composition Determination of Ferrous Oxalate

AIMS

1. To learn how to prepare ferrous oxalate from ammonium ferrous sulfate and to determine its chemical formula.

2. To understand the method of determining the content of iron and oxalate by potassium permanganate.

INTRODUCTION

Ferrous oxalate, or iron(II) oxalate, is a inorganic compound with the formula $FeC_2O_4 \cdot (H_2O)x$ where x is typically 2. These are orange compounds, poorly soluble in water. The dihydrate $FeC_2O_4 \cdot (H_2O)_2$ is a coordination polymer, consisting of chains of oxalate-bridged ferrous centers, each with two aquo ligands. When heated, it dehydrates and decomposes into a mixture of iron oxides and pyrophoric iron metal, with release of carbon dioxide, carbon monoxide, and water.

Ferrous oxalate was formerly used in potassium oxalate solution as a photographic developer. Battery grade ferrous oxalate can be used as a raw material for lithium ironphosphate, which is the battery cathode materials.

Ferrous ammonium sulfate reacts with oxalic acid to form ferrous oxalate solid under appropriate heating conditions:

$$(NH_4)_2SO_4 \cdot FeSO_4 \cdot H_2O + H_2C_2O_4 = FeC_2O_4 \cdot nH_2O + (NH_4)_2SO_4 + H_2SO_4 + H_2O \qquad (2\text{-}19\text{-}1)$$

Owing to the reducibility of Fe^{2+} and $C_2O_4^{2-}$ in the product, it can be determined by redox titration of potassium permanganate. Potassium permanganate, $KMnO_4$, is a strong oxidizing agent. The reduction of permanganate requires strong acidic conditions. Permanganate, MnO_4^-, is an intense dark purple color. Reduction products, Mn^{2+}, is colorless. So, MnO_4^- ions can serve as the self-indicator, and no additional indicator is needed for this titration. A faint pink color means that the end point of the titration is reached.

The titration will be carried out in two steps. First, both Fe^{2+} and $C_2O_4^{2-}$ are titrated with standard potassium permanganate solution, we can get the total molar number of Fe^{2+} and $C_2O_4^{2-}$.

$$5Fe^{2+} + 5C_2O_4^{2-} + 3MnO_4^- + 24H^+ = 5Fe^{3+} + 10CO_2 + 3Mn^{2+} + 12H_2O \quad (2\text{-}19\text{-}2)$$

If you use the total weight of the product substracts the content of Fe^{2+} and $C_2O_4^{2-}$, you'll get the mass of the crystalline water, so Mn before H_2O can be determined.

Second, the oxidation products Fe^{3+} will be reduced with zinc powder, and Fe^{2+} continues to be titrated with standard potassium permanganate solution. At this time, the total molar number of Fe^{2+} can be obtained.

$$5Fe^{2+} + MnO_4^- + 8H^+ = 5Fe^{3+} + Mn^{2+} + 4H_2O \qquad (2\text{-}19\text{-}3)$$

The content of Fe^{2+}, $C_2O_4^{2-}$ and crystalline water can be used to determine the composition of ferrous oxalate.

REAGENTS AND APPARATUS

- Analytical balance, heating mantle with stirrer, stir bars, oven, water pump, 250 mL beakers, 50 mL burettes, 250 mL volumetric flasks, 250 mL conical beakers, 10 mL and 50 mL graduated cylinder, brinell funnel, 500 mL filter bottle, surface dishes, glass rods.
- Ammonium ferrous sulfate ($(NH_4)_2SO_4 \cdot FeSO_4 \cdot 6H_2O$), 1 and 2 mol $\cdot$ L^{-1} sulphuric acid (H_2SO_4) solution, 1 mol $\cdot$ L^{-1} oxalate ($H_2C_2O_4$) solution, 0.02 mol $\cdot$ L^{-1} potassium permanganate ($KMnO_4$) solution, zinc (Zn) powder, 0.1 mol $\cdot$ L^{-1} potassium thiocyanate ($KSCN$) solution, acetone, distilled water.

PROCEDURES

1. Preparation of ferrous oxalate

Weight 6.0 g $(NH_4)_2SO_4 \cdot FeSO_4 \cdot 6H_2O$ in a 250 mL beaker, and record the mass to 4 decimal places. Add 30 mL distilled water to dissolve the solid. Then, add 2 mL 2 mol $\cdot$ L^{-1} H_2SO_4 to make the solution acidity. Heat while stirring to make it dissolve completely. Add 40 mL 1 mol $\cdot$ L^{-1} $H_2C_2O_4$ solution to the beaker. Heat and stir the solution to boiling. After large yellow solid precipitates appear, stop heating and stirring, and let it stand still.

Pour out the supernatant. Be careful not to disturb the solution violently. Add 40 mL distilled water. Heat and stir to near boiling, and let the precipitate to be washed thoroughly. Filter after cooling to room temperature, and wash the solid with a little acetone twice. Put the solid in a surface dish, and spread it flat. Place the surface dish in the oven, and dry at 40 ℃ for 60 mins.

Weigh the product, and record the mass to 4 decimal places.

2. Composition determination of ferrous oxalate

Accurately weight 0.12—0.14 g ferrous oxalate, and record the mass to 4 decimal places. Put the ferrous oxalate in a 250 mL conical beaker. Add 25 mL 2 mol $\cdot$ L^{-1} H_2SO_4 solution, and heat it to 40—50 ℃ to dissolve the sample.

The burette is washed with water and distilled water three times, and rinsed with the 0.02 mol $\cdot$ L^{-1} $KMnO_4$ standard solution, and then filled with 0.02 mol $\cdot$ L^{-1} $KMnO_4$ standard solution. Record the initial reading V_0 to 2 decimal places.

Titration with $KMnO_4$ solution, the colour of solution changes from colorless to yellow-green, then finally to pale pink. The endpoint is reached if the pale pink does not fade in 30 seconds. Record reading V_1 to 2 decimal places.

Add 2 g Zn powder and 5 mL 2 mol $\cdot$ L^{-1} H$_2$SO$_4$ to the conical beaker. Boil for 10 minutes. Use KSCN solution to test the droplets on the drop board. If KSCN solution does not turn red immediately, proceed to the next step. If it turns red immediately, continue boiling for a few more minutes.

Filteration after cooling to room temperature. Wash the conical beaker with 10 mL 1 mol $\cdot$ L^{-1} H$_2$SO$_4$, and still use the 10 mL H$_2$SO$_4$ to wash the precipitate. Transfer all the filtrate, which contains Fe^{2+}, into another conical beaker. Titration with the standard KMnO$_4$ solution until the solution appears pale pink, that is the end point. Read the volume of the consumed KMnO$_4$ V_2 to 2 decimal places.

Individually repeat the titration steps with another two ferrous oxalate samples.

The content of Fe^{2+}, C$_2$O$_4^{2-}$ and crystalline water in the product can be deduced from the titration results. Find out the chemical formula of the product.

DATA TREATMENT

1. Preparation of ferrous oxalate

	reactant: (NH$_4$)$_2$SO$_4$ $\cdot$ FeSO$_4$ $\cdot$ 6H$_2$O	product: FeC$_2$O$_4$ $\cdot$ nH$_2$O	yield
weight			

2. Composition determination of ferrous oxalate

		1	2	3
FeC$_2$O$_4$ $\cdot$ nH$_2$O weighting	m(FeC$_2$O$_4$ $\cdot$ nH$_2$O)/ g			
KMnO$_4$ standard titrant	before titration V_0/mL			
	after 1st titration V_1/mL			
	net volume ΔV_1/mL L			
	after 2^{ndt} titration ΔV_2/mL			
	net volume ΔV_2/mL			
The total content of Fe^{2+} and C$_2$O$_4^{2-}$ /mol				
average content of Fe^{2+} and C$_2$O$_4^{2-}$ /mol				
The content of Fe^{2+} /mol				
averagecontent of Fe^{2+} /mol				
The weight of nH$_2$O/g				
The number of "n"				
average "n"				
Mole ratio Fe^{2+} : C$_2$O$_4^{2-}$: H$_2$O				

QUESTIONS

1. Why is sulfuric acid used to acidify the solution both in preparation and in titration?

2. What is the purpose of adding zinc powder? How to remove excessive zinc powder?

3. In the second step of titration, is there oxalate in the solution? If so, how will it affect the results?

2.20 The Preparation and Analysis of Calcium Iodate Ca(IO$_3$)$_2$

> **AIMS**
>
> 1. To master redox chemistry, especially the oxidation of elemental iodine to iodate.
>
> 2. To learn how to prepare the product Ca(IO$_3$)$_2$ and calculate the yield and then analyse its purity.
>
> 3. To review the principle and procedure of titration analysis.

INTRODUCTION

A kind of chemical reaction is the redox reaction. Redox stands for reduction and oxidation. The original definition of oxidation is "a reaction in which a substance combines with oxygen", while reduction is "a reaction in which oxygen is removed from a substance". In this experiment you are going to prepare a sample of calcium iodate, calculate the yield and then analyse its purity. The synthesis involves the oxidation of elemental iodine to iodate (IO$_3^-$) and the difference of solubility.

Iodine is the heaviest of the stable halogens, it exists as a lustrous, purple-black metallic solid at standard conditions that sublimes readily to form a violet gas. It was discovered by Bernard Courtois in 1811 in France. It is the least abundant of the stable halogens, being the six tieth most abundant element.

Iodine is the fourth halogen, being a member of group 17 in the periodic table, below fluorine, chlorine, and bromine; it is the heaviest stable member of its group. Iodine has an electron configuration of $[Kr]4d^{10}5s^25p^5$, with the seven electrons in the fifth and outermost shell being its valence electrons. Like the other halogens, it is one electron short of a full octet and is hence a strong oxidizing agent, reacting with many elements in order to complete its outer shell, although in keeping with periodic trends, it is the weakest oxidizing agent among the stable halogens: it has the lowest electronegativity among them, just 2. 66 on the Pauling scale (compare fluorine, chlorine, and bromine at 3. 98, 3. 16, and 2. 96 respectively; astatine continues the trend with an electronegativity of 2. 2). Elemental iodine hence forms diatomic molecules with chemical formula I_2, where two iodine atoms share a pair of electrons in order to each achieve a stable octet for themselves; at high temperatures, these diatomic molecules reversibly dissociate a pair of iodine atoms. Similarly, the iodide anion, I^-, is the strongest reducing agent among the stable halogens, being the most easily oxidized back to diatomic I_2.

Iodine occurs in many oxidation states, including iodide (IO^-), iodate (IO_3^-), and the various periodate anions (i. e. IO_4^-). Iodine oxides are the most stable of all the halogen oxides, because of the strong I-O bonds resulting from the large electronegativity difference between iodine and oxygen. Iodates are by far the most important of these compounds, which can be made by oxidizing alkali metal iodides with oxygen at 600 ℃ and high pressure, or by oxidizing iodine with chlorates. Iodates are stable to disproportionation in both acidic and alkaline solutions. From these, salts of most metals can be obtained.

In addition, the difference of solubility is widely used to prepare compounds. In this experiment, $Ca(IO_3)_2$ is slightly soluble in water, but KIO_3 is easily soluble, so you can obtain the product $Ca(IO_3)_2$ from KIO_3 solution.

REAGENTS AND APPARATUS

- Analytical balance, heating mantle, oven, water pump, 100 mL beaker, 50 mL burettes, 250 mL conical beakers, 10 mL and 50 mL graduated cylinder, 500 mL filter bottle, Buchner funnel, watch glass, glass rods.
- Potassium chlorate ($KClO_3$), 6 mol • L^{-1} hydrochloric acid (HCl), 30% potassium hydroxide (KOH), 0. 1 mol • L^{-1} sodium hydroxide (NaOH), 1 mol • L^{-1} calcium chloride ($CaCl_2$), 1 : 1 (volume ratio) perchloric acid ($HClO_4$), potassium iodide (KI), 0. 1 mol • L^{-1} Sodium thiosulfate ($Na_2S_2O_3$), 0. 5% starch solution, distilled water.

PROCEDURES

1. Preparation of Ca(IO₃)₂

Transfer 2.00 g KClO₃ (accurate to is 0.1 mg) into a 100 mL round-bottom flask containing 45 mL of hot water. Remove the flask from the source of heat and immediately add 2.20 g powdered iodine. Without delay slowly add 8 drops 6 mol·L⁻¹ HCl with stirring to control the pH to 1. Install the condenser, and use a rubber tube at the top of the condenser to introduce the generated gas into a beaker containing NaOH solution. Keep 85 ℃ by water bath until the iodine has been consumed, and the solution turns to pale yellow.

Add 30% KOH to the solution and control the pH to 10. Then add 1 mol·L⁻¹ CaCl₂ dropwise, and keep swirling the solution. Cool to room temperature, and then continue to cool in ice-water bath. Filter off the precipitate using a Buchner funnel and wash the product with 2 mL ice-cold water 3 times. Transfer the filter paper and product to a watch glass and dry for about 1 hour at 60 ℃.

Record the yield and the appearance of your product and calculate the theoretical yield of Ca(IO₃)₂. Write an equation for the overall reaction.

2. Analysis of Ca(IO₃)₂

Weigh accurately about 0.6 g Ca(IO₃)₂ (accurate to 0.1 mg) into a 100 mL beaker. Add 20 mL HClO₄ (1 : 1 by volume), and dissolve the sample with gentle heat. After cooling, transfer it to a 250 mL volumetric flask, and fill the volume with distilled water. Transfer 50.00 mL solution into 250 mL iodine flask, add 2 mL HClO₄ (1 : 1 by volume) and 3 g KI. Cap the iodine flask, swirl it to let KI dissolve and leave it in the dark for 3 minutes.

Add 50 mL distilled water into the iodine flask, titrate the solution to pale straw using 0.1 mol·L⁻¹ Na₂S₂O₃. Add 2 mL 0.5% starch solution, continue titrate with 0.1 mol·L⁻¹ Na₂S₂O₃ until blue disappears. You need to repeat twice again.

Calculate the percentage IO₃⁻ in your compound and compare this with the expected value.

DATA TREATMENT

1. Preparation of Ca(IO₃)₂

	reactant: KClO₃	reactant: I₂	product: Ca(IO₃)₂	yield
weight				

2. Analysis of $Ca(IO_3)_2$

		1	2	3
$Ca(IO_3)_2$ weighting	$m(Ca(IO_3)_2)/g$			
$Na_2S_2O_3$ titrant	beforetitration/mL			
	after titration/mL			
	net volume/mL			
$n(Ca(IO_3)_2)/mol$				
the percentage of IO_3 %				
average percentage %				
the average relative deviation				

QUESTIONS

1. Write the half equation of the oxidation reaction and the reduction reaction, and the balanced stoichiometric equation for the preparation and analysis of $Ca(IO_3)_2$.

2. How to prepare $0.1\ mol \cdot L^{-1}\ Na_2S_2O_3$ standard solution?

3. Why should starch be added at the later stage of titration, that is, when the color of the solution becomes pale straw?

2.21 Preparation and Zinc Content Determination of Zinc Gluconate

AIMS

1. To learn and master how to prepare zinc gluconate.
2. To learn the determination method of zinc content in zinc salt.

INTRODUCTION

Zinc is a naturally occurring mineral. Zinc is important for growth and for the development and health of body tissues. Zinc can activate a variety of important antioxidant enzymes, thereby eliminating the damage of oxygen free radicals, and maintaining the normal permeability of the cell membrane to protect the normal biochemical composition of cell membranes, metabolic structure and function. Zinc can not only induce the activation of T lymphocyte but also activate the B lymphocytes. Zinc is also involved in antibody formation and release, and stimulating immune cells to secrete a variety of cytokines. Lack of zinc in elderly people can cause immune function disorder; zinc can affect the synthesis of insulin, secretion, storage, degradation and biological activity, being the major trace amount element directly affecting the insulin physiology. Zinc can improve the body's sensitivity to insulin.

Zinc gluconate is the zinc salt of gluconic acid. The chemical formula of this compound is $C_{12}H_{22}O_{14}Zn$. It has the molar mass, $455.68 \text{ g} \cdot \text{mol}^{-1}$. In its pure form, zinc gluconate is a white to off-white powder. Zinc gluconate is soluble in water, soluble in boiling water, and insoluble in absolute ethanol, chloroform and ether. It contains one zinc cation (Zn^{2+}) associated with two anions of gluconic acid. The structure is shown in Fig. 1.

Moreover, zinc gluconate is a popular dietary supplement and a good source of zinc. Zinc gluconate has fewer side effects than some inorganic zinc like zinc sulfate, and better absorption as well as a zinc supplement. It has certain efficacy on the treatment of zinc deficiency caused growth retardation, malnutrition, anorexia, recurrent oral ulcers, acne and senile zinc deficiency as well as immune dysfunction. Zinc gluconate may also be used for other purposes not listed in this medication guide.

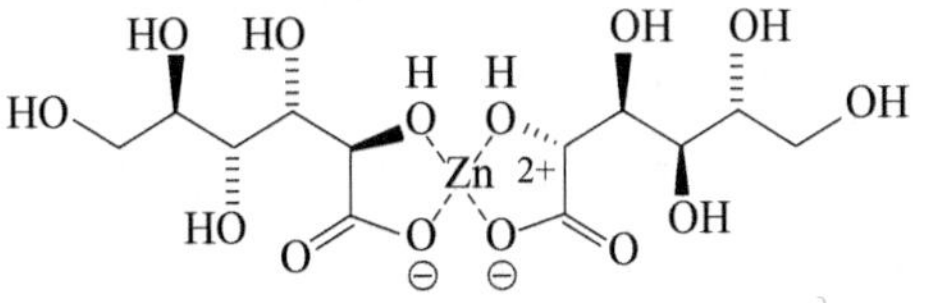

Figure 2 - 25　Zinc Gluconate Structure

Zinc gluconate can be prepared directly from gluconic acid and zinc oxide or salt. In this experiment, zinc gluconate is prepared by direct reaction of zinc sulfate with calcium gluconate. The reaction equation is as follows:

$$Ca(C_6H_{11}O_7)_2 + ZnSO_4 \Longrightarrow Zn(C_6H_{11}O_7)_2 + CaSO_4 \qquad (2\text{-}21\text{-}1)$$

$CaSO_4$ is then removed by filtration, and the solution is concentrated and re-crystallized to form colorless or white zinc gluconate crystals.

Zinc gluconate should be tested for several items before it is used as a supplement. This experiment is only a preliminary analysis of product quality. Zinc content in the product is determined by EDTA coordination titration.

Zinc content in the zinc gluconate can be calculated by the following equation:

$$Zn\% = \frac{c_{Na_2 EDTA} \times V_{Na_2 EDTA} \times 65}{m} \times 100\% \qquad (2\text{-}21\text{-}2)$$

Where $c_{Na_2 EDTA}$ is the concentration of standard solution, $V_{Na_2 EDTA}$ is the volume of standard solution, m is the mass of zinc gluconate.

REAGENTS AND APPARATUS

- Analytical balance, heating mantle, water bath, oven, water pump, 100 mL and 200 mL beakers, 50 mL burettes, 250 mL conical beakers, 10 mL and 50 mL graduated cylinder, 500 mL filter bottle, Buchner funnel, evaporating dish, watch glass, glass rods, filter paper.
- Zinc sulfate heptahydrate ($ZnSO_4 \cdot 7H_2O$), calcium gluconate ($Ca(C_6H_{11}O_7)_2$), 0.02 mol $\cdot$ L^{-1} disodium ethylenediamine tetraacetic (Na_2EDTA) solution, 0.1 mol $\cdot$ L^{-1} NH_3/NH_4^+ buffer (pH=10), 5 g $\cdot$ L^{-1} Eriochrome black T solution, 95% ethanol, distilled water.

PROCEDURES

1. Preparation of zinc gluconate

1.1 Preparation of crude product

Add 40 mL distilled water to a 200 mL beaker, and heat the beaker in water bath to 80—90 ℃. Weight 6.7 g $ZnSO_4 \cdot 7H_2O$ and 10 g calcium gluconate ($Ca(C_6H_{11}O_7)_2$) and record the mass to 4 decimal places. Carefully add $ZnSO_4 \cdot 7H_2O$ to the hot water, and stir with glass rod until the salt is completely dissolved. Then place the beaker in a water bath at 90 ℃, gradually add calcium gluconate. Be careful. Stir until calcium gluconate is completely dissolved, and keep it in the water bath for 20 minutes.

Vaccum filtration while the solution is hot. The solid in the Brinell funnel is $CaSO_4$ and we do not need it. The filtrate is then transferred to a 200 mL beaker and heated to near boiling with a heating mantle. 0.1 g activated carbon is added to decolorize the solution. The hot solution is then filtered under vaccum conditions.

The filtrate is transferred into an evaporating dish and slowly heat to concentrate the solution until it is viscous. Cooling the solution to room temperature. Add 95% ethanol 20 mL to reduce the solubility of zinc gluconate, and stir constantly. At this time, a large number of colloidal zinc gluconate precipitate. After fully stirring, use the dumping method to remove ethanol. Adding another 20 mL 95% ethanol to the colloidal precipitation. After fully stirring, the crude zinc gluconate is obtained by vaccum filtering the colloidal precipitation to dry. Mother liquor should be recovered.

1.2 Re-crystallization

Add 10 mL distilled water to a 100 mL beaker, and heat the beaker in water bath to 90 ℃. Add the crude zinc gluconate product, stir until dissolved, and then vacuum filter when the solution is hot. The filtrate is cooled to room temperature. Add 10 mL 95% ethanol while stirring. After crystallization appears, vacuum filter and allow the solid to dry. The pure zinc gluconate is obtained after drying in a oven at 50 ℃ for 1 hour. Finally, weigh the pure product and record the mass to 4 decimal places, and calculate the yield.

2. Determination of zinc content in products

Weight 0.46 g pure $Zn(C_6H_{11}O_7)_2$ and record the mass to 4 decimal places. Place it in the conical beaker. Add 50 mL distilled water, slightly heat to let $Zn(C_6H_{11}O_7)_2$ dissolve completely. Add 5 mL NH_3-NH_4Cl buffer (pH = 10.0) and 3—4 drops chrome black T indicator to the conical beaker.

Wash a 50.00 mL burette thoroughly with water, and then rinse it with distilled water and a few mLs of Na_2EDTA solution sequently. Drain through the stopcock and then fill the burette with 0.02 mol · L^{-1} Na_2EDTA solution. When the solution change from wine red to blue, the endpoint is reached. Repeat three times in parallel and calculate the zinc content.

DATA TREATMENT

1. Preparation of zinc gluconate

	reactant: $ZnSO_4 · 7H_2O$	reactant: $Ca(C_6H_{11}O_7)_2$	product: $Zn(C_6H_{11}O_7)_2$	yield
weight				

2. Determination of zinc content in products

		1	2	3
$Zn(C_6H_{11}O_7)_2$ weighting	$m(Zn(C_6H_{11}O_7)_2)/g$			
Na_2EDTA titrant	before titration/mL			
	after titration/mL			
	net volume/mL			
Zn content $w\%$				
AverageZn content $w\%$				
the average relative deviation				

QUESTIONS

1. Can zinc gluconate be prepared by the reaction of zinc chloride or zinc carbonate with calcium gluconate? Explain.

2. If the zinc content calculated is not consistent with the theoretical value, what may be the reasons?

2.22 Synthesis and Characterization of Geometrical Isomers of Dichlorobis(ethylenediamine)- Cobalt (III) Chloride

AIMS

1. Synthesis of *trans*-dichlorobis (ethylenediamine) cobalt (III) chloride.

2. Conversion of the trans to the cisform.

3. Understanding the principles of stereo-isomerism in co-ordination complexes.

INTODUCTION

Co-ordination compounds with same formula but different arrangements of atoms are called isomers. The study of isomerisms is important in co-ordination chemistry as different isomers have different physical and chemical properties. Isomers are classified as structural and stereo-isomers. The stereo-isomers may be further categorized as geometric and optical isomers. This experiment aims at the synthesis of geometric isomers of the co-ordination complex dichloro bis (ethylenediamine) cobalt (III) chloride.

The first step (2-22-1) is the synthesis of *trans*-dichloro bis (ethylenediamine) cobalt (III) chloride from cobalt (II) chloride hexahydrate and ethylenediamine.

$$CoCl_2 \cdot 6H_2O + H_2N \diagdown\diagup NH_2 + HCl + O_2 \longrightarrow \begin{bmatrix} & Cl & \\ H_2N & | & NH_2 \\ & Co & \\ N & | & N \\ H_2 & Cl & H_2 \end{bmatrix} Cl \cdot HCl$$

(2-22-1)

The second step (2-22-2) is the conversion of the *trans* form into the *cis* form.

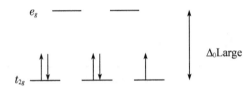 (2-22-2)

In this experiment we will focus on the synthesis and characterization of the *cis* and *trans* isomers. Based on the structure of the geometric isomers, the possible optical activity of the geometric isomers will also be predicted. For advanced levels of inorganic chemistry experiments, optical studies may be done for the detailed analysis of the optical isomers.

Co^{3+} has a low-spin $3d_6$ electronic configuration $(1s^2 2s^2 2p^6 3s^2 3p^6 3d^6)$, shown in Figure 2 – 26. Use a spectrophotometer to record their UV-visible spectra in aqueous solution. From the λ_{max} value recorded, the vital quantity Δ_o, the energy splitting of the t_{2g} and e_g subset-orbitals, can easily be calculated.

Figure 2 – 26 low-spin states of $3d_6$ electrons

REAGENTS AND APPARATUS

- Analytical balance, heating mantle with stirrer, stir bars, 250 mL and 100 mL beakers, surface dishes, glass rods, spatula, steam bath, Buchner funnel and vacuum filtration unit.
- Cobalt (II) chloride hexahydrate, ethanol, methanol, ethylenediamine, diethylether, conc. Hydrochloric acid.

PROCEDURES

1. The synthesis of *trans*-dichloro bis (ethylenediamine) cobalt (III) chloride

Add 1. 6 g cobalt (II) chloride hexahydrate to 5 mL water. To this add 6 g ethylenediamine and stir continuously. Through this solution pass a vigorous stream of air for 10—12 hours. Then add 3. 5 mL conc. hydrochloric acid to this solution and evaporate it on a steam bath so that the volume of the solution reduces to around

10 mL. Cool the solution to room temperature and let it stand overnight. Filter the bright green precipitate and wash it with ethanol and diethylether. Dry the product at 110 ℃. At this drying step, the HCl will be lost and the bright green crystals will be transformed into a dull green powder. Weigh the product.

2. Conversion of the *trans* isomer to the *cis* form

Evaporate a solution of *trans*-dichloro bis (ethylenediamine) cobalt (III) chloride to dryness in a steam bath. Wash with cold water to remove the unchanged *trans* form. For complete conversion of *trans* to *cis* form the evaporation step may be repeated two or three times. However further repetition may lead to decomposition of the complex. Weigh the product.

3. Obtain the IR spectra of *trans* and *cis* isomers. Obtain the UV-Vis spectra of *trans* and *cis* isomers in methanol.

DATA TREATMENT

	trans complexe	*cis* complex
Starting material		
Mass taken (g)		
Formula weight of Product(g • mol^{-1})		
Color of product		
Predicted mass (g)		
Actual mass (g)		
Yield %		
λ_{max} (nm)		
Δ_o (kJ • mol^{-1})		

QUESTIONS

1. Compare the IR spectra of both *trans* and *cis*-complexes. Assign peaks and characterize and interpret the important aspects of the spectra.

2. Compare the UV-Vis spectra of both *trans* and *cis*-complexes. Assign peaks and characterize and interpret the important aspects of the spectra.

3. Draw the structures of both *trans* and *cis*-isomers and determine the symmetry point groups of each.

2.23 Synthesis and Characterization of The Complex bis (*N*, *N′*-Disalicylalethylenediamine)-μ-Aquadicobalt (II)

> **AIMS**
>
> 1. Synthesis of the ligand *N*, *N′*-disalicylidene ethylenediamine.
>
> 2. Synthesis of the complex bis (*N*, *N′*-disalicylalethylenediamine)-μ-aquadicobalt (II).
>
> 3. Understanding the structure and geometry of this chelated complex.

INTRODUCTION

In co-ordination chemistry the ligands attached to the metal center can be classified according to their denticity. The "claw-like" attachment of the multi-dentate ligands to the metal center gives rise to the term "chelate". The formation of a chelated complex provides additional thermodynamic stability to the co-ordination complex due to a greater degree of association between the metal and the ligands.

Biologically active metal centers chelated with multi-dentate ligands have been found to be useful in serving as model complexes in bio-inorganic chemistry. This experiment aims at the synthesis of a cobalt (II) chelated complex that serves as a model for studying reversible co-ordination of oxygen in oxygen transport systems present in higher organisms.

The first step is the synthesis of the multi-dentate ligand *N*, *N′*-disalicylidene ethylenediamine. The ligand can be synthesized from salicylaldehyde and ethylenediamine as the following reaction (2-23-1):

$$ (2\text{-}23\text{-}1) $$

The second step (2-23-2) is the synthesis of a dimeric co-ordination complex containing the above ligand and two cobalt (II) centers. The reaction is shown below:

$$+ \text{CoCl}_2 \xrightarrow{\text{Base}} \qquad (2\text{-}23\text{-}2)$$

In this experiment we will focus on the synthesis and characterization of this dimeric complex. For advanced levels of inorganic chemistry experiments, oxygen uptake studies may be done for the detailed analysis of the active and inactive forms of cobalt (II) complexes containing the N, N'-disalicylidene ethylenediamine ligand, that participate in reversible oxygen binding reactions.

REAGENTS AND APPARATUS

- Analytical balance, melting point apparatus, heating mantle with stirrer, stir bars, 250 mL and 100 mL beakers, 250 mL round bottom flasks, surface dishes, glass rods, spatula, centrifuge tubes, centrifuge, Buchner funnel and vacuum filtration unit.
- Salicylaldehyde, ethylenediamine, ethanol, methanol, sodium hydroxide, sodium acetate trihydrate, cobalt (II) chloride hexahydrate.

PROCEDURES

1. The synthesis of the ligand N, N'-disalicylidene ethylenediamine:

Solution A: In a round bottom flask, take 25 mL of ethanol and dissolve 3.05 g salicylaldehyde in it.

Solution B: In a small beaker take 3 mL ethanol and dissolve 0.75 g ethylenediamine in it.

Heat solution A to 70 ℃ and add solution B to it. The color of the resulting solution turns yellow. Heat this solution for 30 minutes under reflux. By evaporating the solvent, reduce the volume of the solution to half. Cool the solution to room temperature and collect the yellow crystalline precipitate by filtration under suction. Wash the precipitate with ice-cold ethanol. Recrystallize from methanol and dry the final product under vacuum suction at room temperature. Weigh the product.

2. The synthesis of the complex bis (N, N'-disalicylalethylenediamine)-μ-aquadicobalt (II):

Solution A: In a beaker take 150 mL water and dissolve 1.34 g finely ground *ligand N, N'*-disalicylidene ethylenediamine in it. Now add 3.9 g sodium hydroxide and 5 g sodium acetate trihydrate to it and continuously stir for 10—15 minutes.

Solution B: In a small beaker take 25 mL of hot water and dissolve 1.23 g cobalt (II)chloride hexahydrate to it.

Add solution B to solution A and keep on stirring until it forms into a reddish brown paste. Now let this mixture stand for 15 minutes at room temperature. Centrifuge the mixture to remove the mother liquor and collect the hard cake-like precipitate. Wash the precipitate three times with 10 mL of water each. Then remove the precipitate from the centrifuge tube and add 75 mL water to it. Mix thoroughly such that all lumps are removed and uniform slurry is obtained. Now centrifuge the slurry and collect the final product. Break up the final product into small pieces and dry at 100 ℃ under reduced pressure. Weigh the product.

3. Determine melting points of the ligand *N, N'*-disalicylidene ethylenediamine and the complex bis(*N, N'*-disalicylalethylenediamine)-μ-aquadicobalt (II). Obtain the IR spectra of them.

DATA TREATMENT

	ligand	complex
Color of product		
Predicted mass (g)		
Actual mass (g)		
Yield %		
melting points (℃)		

QUESTIONS

1. For the ligand and the complex, analyze and compare the IRspectra obtained by assigning the peaks and interpreting the important aspects of the spectra.

2. The structure of the dicobalt complex you have prepared is given in the experimental section. Draw the structure of the possible *monomeric* form of this dicobalt complex and name the geometry.

3. Determine the spin state of the *monomeric* form, write the electron configuration of the cobalt center in the monomer, and draw the relevant fully labeled crystal field splitting diagram.

2.24 Spectrophotometric Determination of Iron

AIMS

1. To learn how to draw the standard curve and determine the sample content in spectrophotometric method.

2. To learn the principles, components and usage of spectrophotometer.

INTRODUCTION

Iron is the fourth most abundant element in the earth's crust and is an important component in many biological systems. The fact that iron has two readily accessible oxidation states, Fe^{2+} and Fe^{3+}. The ability to measure the concentration of iron in aqueous solutions in a quick and efficient way is important to many industries. Manufacturing industries where metal parts need to be cleaned may need to determine the level of iron in their waste streams for environmental compliance. Governments at all levels have an interest in testing wastewater, natural waters, and drinking waters to determine iron content to ensure compliance with the law and to ensure the safety of the water supply for wildlife and the human population.

Well-equipped, modern laboratories may perform iron content analysis by atomic emission spectroscopy in an inductively coupled plasma (ICP) spectrometer. Flame atomic absorption spectrometry also can be used, but iron solutions are notorious for clogging the burner with iron oxide when the concentration exceeds a certain level. Both of these techniques require a significant investment in instrumentation and a sustained laboratory infrastructure involving compressed gases and control of exhaust vapors. Fortunately, the solution chemistry underlying a colorimetric determination of iron content is simple enough to be reduced to kit form and performed in the field with hand-held equipment or in the lab with a low-cost visible spectrophotometer and simple glassware.

Colorimetric analysis is based on the change in the intensity of the colour of a solution with variations in concentration. Colorimetric methods represent the simplest form of absorption analysis. The human eye can even be used as the detector to compare the colour of the sample solution with a set of standards until a match is found. The colorimetric determination of iron content involves the measurement of the ferrous ion(Fe^{2+}) when it forms a complex with three molecules of 1, 10-phenanthroline, also called ortho-phenanthroline or abbreviated as phen (Figure 2 – 27). The complex formed by one Fe^{2+} ion and three 1, 10-phenanthroline molecules, ferrous tris (1, 10-phenanthroline)iron(II) or $[Fe(phen)_3]^{2+}$, is a bright orange-red color. A 3D model of the structure of the complex is shown in Figure 2 – 28. The absorption spectrum of the complex has a maximum at about 510 nm. This complex is stable indefinitely at pH values of 3 or higher.

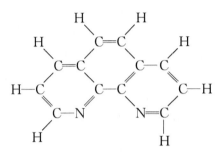

Figure 2 – 27 The structure of 1, 10-phenanthroline.

Figure 2 – 28 3D model of the $[Fe(phen)_3]^{2+}$ complex.

Upon adding the phenanthroline, reactionoccurs:

$$Fe^{2+} + 3phen = [Fe(phen)_3]^{2+} \qquad (2\text{-}24\text{-}1)$$

In addition to its Fe^{2+} (ferrous ion) form, iron also can exist in a Fe^{3+} (ferric ion) form. To perform a total iron measurement, it is essential to reduce any Fe^{3+} in solution to Fe^{2+} before adding the phenanthroline to form the complex. The chosen reducing agent in this experimental protocol is hydroxylamine hydrochloride. The reaction depends on whether the solution is acidic or basic.

Acid：　　　$2Fe^{3+} + 2NH_3OH^+ \Longrightarrow 2Fe^{2+} + N_2 + 4H^+ + 2H_2O$　　　　(2-24-2)

Base：　　　$2Fe^{3+} + 2NH_2OH + 2OH^- \Longrightarrow 2Fe^{2+} + N_2 + 4H_2O$　　　　(2-24-3)

Sodium acetate is added to the solution to adjust the acidity to a level at which the iron(II)-phenanthroline complex is especially stable.

To determine the concentration of iron in an unknown solution, we must first calibrate the method with the spectrophotometer using Beer's Law：

$$A = \varepsilon bc \qquad (2\text{-}24\text{-}4)$$

where A is the absorbance reported by the spectrophotometer; ε is the molar absorptivity, a value that describes how strongly the particular compound absorbs photons at the particular wavelength, typically with units of $(L \cdot cm^{-1} \cdot mol^{-1})$; b is the pathlength of the cuvette in cm, where typically a 1 cm pathlength cuvette is used; c is the concentration of the solution in $mol \cdot L^{-1}$.

the absorbance of each solution on the y-axis versus concentration on the x-axis：

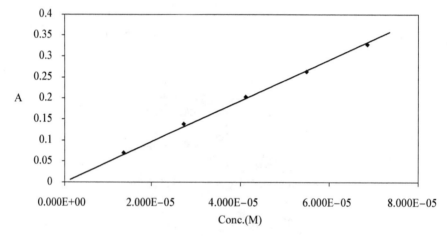

Figure 2 - 29　Example of a generic Beer's Law Plot

Comparing the equation for Beer's Law to the plot, we see that the slope of the line is equal to ε. We can use the plot to calculate the concentration of an unknown solution in one of two ways：

1. Use the plot itself to select the point on the y-axis representing the measured absorbance. Trace a horizontal line from that point to the plot line, then draw a vertical line straight down to the concentration on the x-axis. The point where this intersects the x-axis represents the concentration of the unknown solution.

2. Use the equation of the line. If $A = \varepsilon bc$ and $b = 1$, then $c = A/\varepsilon$. Use a spreadsheet program or a graphing calculator to plot your data and determine a best-fit line (trend line) to calculate the slope of your line. This slope equals ε. Divide the measured A value for the unknown by ε to calculate the concentration of the unknown solution.

In this experiment, you will perform an analysis of an iron-containing solution with an unknown concentration. To determine the concentration of iron(II), you will first measure the absorbance of a fixed wavelength of light by standard solutions containing known concentrations of the iron(II) ion. You will use these data to prepare a calibration graph showing absorbance as a function of concentration. You will then measure the absorbance at the same wavelength of light by the "unknown" solution. Using the calibration curve, you will be able to determine the concentration of the total iron and iron(II) in the unknown.

REAGENTS AND APPARATUS

- Spectrophotometer, cuvettes, analytical balance, 50 mL and 250 mL volumetric flasks, 10 mL measuring pipettes, 25 mL transfer pipettes, 100 mL beakers, 10 mL graduated cylinder, dropper, glass rods, tissue.
- 10 mg $\cdot$ L^{-1} ammonium ferric sulfate ($NH_4Fe(SO_4)_2$) standard solution, 1 mol $\cdot$ L^{-1} sodium acetate (NaAc) solution, 0.5 mol $\cdot$ L^{-1} hydroxylamine hydrochloride ($NH_2OH \cdot HCl$), 0.4% phenanthroline ($C_{12}H_8N_2$) solution, unknown iron solution, distilled water.

PROCEDURES

1. Preparation of 10 mg $\cdot$ L^{-1} Fe^{3+} standard solution

Pipette 25 mL 100 mg $\cdot$ L^{-1} Fe^{3+} stock solution in a 250 mL volumetric flask. Dilute with distilled water to the 250 mL line. Stopper each flask, then invert repeatedly for 2 minutes to let the solution mix completely.

You must not contaminate or dilute the stock solutions everyone in the lab is relying on! When pipetting, pour out small portions of the reagent you need from the stock bottles into a beaker. Never insert a pipette or other glassware into a stock bottle, and never pour unused reagents back into stock bottles.

2. Preparation of the "six" known Fe^{3+} standard solution

Set up six 50 mL volumetric flasks, and they should be emptied and must be chemically clean. Label them as 1, 2, 3, 4, 5, and 6. Pipette 10 mg $\cdot$ L^{-1} Fe^{3+} standard solution into them successively as follows: 0, 2, 4, 6, 8, and 10 mL. Then, 5 mL 1 mol $\cdot$ L^{-1} sodium acetate, and 1 mL 0.5 mol $\cdot$ L^{-1} hydroxylamine hydrochloride are added successively into each flask, and wait 5 to 10 minutes for the reduction of any Fe^{3+} to be completed. Then add 2 mL 0.4% phenanthroline into each flask. Dilute with water to the mark and mix thoroughly by inverting the stoppered flask for 2 minutes. Allow 10 minutes for the reddish-orange color to develop completely.

3. Prepare the Beer's Law plot

Pipet 3 mL of the solution from Flask 1 into a cuvette. This solution will be your blank. Wipe the outside faces of the cuvette with a laboratory tissue and place the cuvette into the square cuvette stage of the spectrophotometer sample compartment with the clear faces pointing to the left and right. Close the lid of the spectrophotometer. Set up a scan from 400 nm to 900 nm in ABS mode and press the 0. 00 button to record a baseline. Wait until the scan has completed. Open the lid, remove the cuvette with the blank solution, and set it aside. If you only have one cuvette, discard the blank solution and rinse the cuvette with deionized water before reusing it.

Pipet 3 mL of the solution from Flask 6 into a cuvette. Wipe the outside faces of the cuvette with a laboratory tissue and place the cuvette into the square cuvette stage of the spectrophotometer sample compartment with the clear faces pointing to the left and right. Close the lid and press Start button to record the scan. When the scan appears on the screen, mark the highest point on the peak. Record the wavelength of maximum absorption, known as λ_{max}.

Set the spectrophotometer to Live Display Mode with measurements in Absorbance at λ_{max}. Follow the same directions for filling, wiping and orienting a cuvette mentioned above, using a cuvette with a blank solution from Flask 1 to record a blank value. Prepare and measure the absorbance of cuvettes containing the solutions from Flasks 2 through 6. Record the values in your Lab Report.

4. Determine the total iron and iron(II) concentration in an unknown sample

Set up two 50 mL volumetric flasks, and they should be emptied and must be chemically clean. Label them as 7 and 8. Pipette 10 mg • L^{-1} unknown solution into them, respectively. Add 5 mL 1 mol • L^{-1} sodium acetate, and 1 mL 0. 5 mol • L^{-1} hydroxylamine hydrochloride in Flask 7, and wait 5 to 10 minutes for the reduction of any Fe^{3+} to be completed. In Flask 8, only add 5 mL 1 mol • L^{-1} sodium acetate, and do not add any hydroxylamine hydrochloride. Then add 2 mL 0. 4% phenanthroline into each flask. Dilute with water to the mark and mix thoroughly by inverting the stoppered flask for 2 minutes. Allow 10 minutes for the reddish-orange color to develop completely.

Record the absorbance of the solution at λ_{max} using the spectrophotometer.

Determine the concentration of total iron in Flask 7 and iron(II) in Flask 8 in the solution mathematically using the Beer's Law equation.

1. Disposal of chemicals

Check with your instructor before discarding any solutions. All solutions can be poured down the sink and rinsed with lots of water to dilute. Discard solids in the trash.

DATA TREATMENT

1. The Beer's Law plot

Flask	1	2	3	4	5	6
$c(Fe^{3+})$ $(mg \cdot L^{-1})$						
absorption A						
Equation $A \sim c$						
$\varepsilon(L \cdot cm^{-1} \cdot mol^{-1})$						
correlation coefficient r						

2. Drawing standard curve

3. Determine the total iron and iron(II) concentration

Flask	7	8
absorption A		
$c(Fe^{3+})$ $(mg \cdot L^{-1})$		

QUESTIONS

1. How to prepare 1 000 mL of 1 000 mg $\cdot$ L^{-1} Fe^{3+} standard solution with ammonium ferric sulfate dodecahydrate ($NH_4Fe(SO_4)_2 \cdot 12H_2O$)? (How many grams of $NH_4Fe(SO_4)_2 \cdot 12H_2O$ should be weighted? What devices or glass wares can be used?)

2. What is the wavelength range of visible light? What is the usage of visible spectrophotometer?

3. What does the molar absorptivity mean? What factors will affect its value?

2.25 Two-component Analysis of A Phenol/Benzoic Acid Mixture by UV-visible Spectroscopy

AIMS

1. To learn the basic structure and usage of ultraviolet-visible spectroscopy.

2. To master the method of qualitative analysis and quantitative analysis with ultraviolet-visible spectroscopy.

INTRODUCTION

UV-visible radiation (in the range of 200—700 nm of the electromagnetic spectrum) interacts with matter by exciting electrons in molecular bonds. In this process, the electrons in chemical bonds are made to "jump" to a higher quantum state by absorption of energy from the radiation. Single bonds require a great deal of energy to undergo electronic excitation and generally do not feature in UV-Vis spectroscopy at all. By contrast, double bonds typically absorb UV radiation at wavelengths around 200 nm and conjugation of a double bond results in the value of the absorbance shifting to longer wavelength. Such extended systems of double bonds are known as "chromophores" and the most common chromophores encountered in organic compounds are associated with benzene rings.

As well as the wavelength at which electromagnetic radiation is absorbed (plotted on the X-axis), the UV spectrum of a sample is also characterized by an absorption, A, plotted on the Y-axis, which is a measure of the amount of light absorbed by the sample at any given wavelength. In order to understand the mathematical definition of absorption, it helps to consider the design of a UV-Vis spectrometer.

The spectrometer incorporates a variable wavelength light source, which can be tuned to emit radiation between 190 and 900 nm. As can be seen from Figure 1, the light passes from the source through the sample cell (as well as a reference cell), and the intensity of the emerging radiation is monitored by a detector situated on the opposite side of these cells. The absorption of UV light by the sample is then defined in the following way (Eqn. 2-25-1):

$$A = \text{Log } I_0/I_t = \varepsilon bc \qquad (2\text{-}25\text{-}1)$$

Where I_0 is the intensity of the radiation from the light source which is incident on the sample; I_t is the intensity of radiation transmitted through the sample; ε is a constant, known as the molar extinction coefficient (obtained from the absorbance of a 1 mol $\cdot$ L^{-1} solution of the sample); b is the pathlength of the cell (expressed in cm; cells are usually 1 cm wide); and c is the concentration of the sample (expressed in moles/litre).

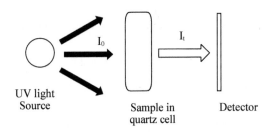

Figure 2 – 30　Design of a UV-VIS spectrometer

When the wavelength of the source in the UV-VIS spectrometer is varied over its full range, a UV spectrum results. For example, the UV spectrum of benzene, shown in Figure 2 – 31, has a principal absorbance at 203. 5 nm (known as the K band), and a weaker absorbance at higher wavelength (referred to as the B band). The UV spectrum of benzene is most easily described by noting the wavelengths at which maxima occur in the spectrum; as well as the absorbtion value associated with each maximum, i. e.

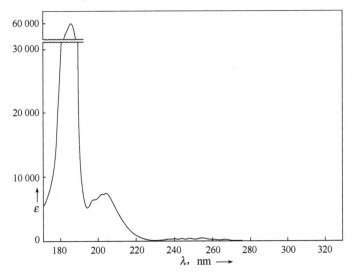

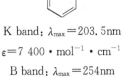

K band: $\lambda_{max} = 203. 5$nm

$\varepsilon = 7\ 400 \cdot$ mol$^{-1} \cdot$ cm^{-1}

B band: $\lambda_{max} = 254$nm

$\varepsilon = 204$ L $\cdot$ mol$^{-1} \cdot$ cm^{-1}

Figure 2 – 31　UV spectrum of benzene

If two compounds, A and B, both have a significant absorbance at the same wavelength, then the absorbance of a mixture of the two components is the sum of each individual contribution, as defined by Eqn. 2-25-2.

$$A = \varepsilon_A bc_A + \varepsilon_B bc_B \qquad (2\text{-}25\text{-}2)$$

In this experiment, both phenol and benzoic acid contain benzene rings, and thus have similar UV spectra.

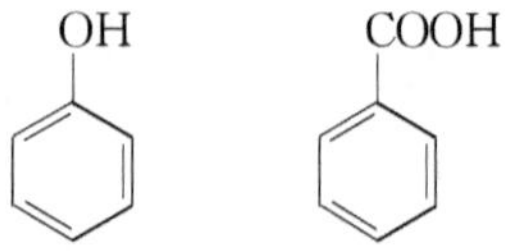

OH   COOH

Figure 2 – 32   Structures of phenol and benzoic acid

By the end of the experiment you should be able to:

1. Operate a UV-spectrometer and measure the UV-visiblespectra of phenol and benzoic acid.

2. Use the results of UV-spectrometric analysis to determine the concentrations of phenol and benzoic acid in a mixture of unknown composition.

REAGENTS AND APPARATUS

- UV-VIS spectrometer, cuvettes, analytical balance, 50 mL and 1 L volumetric flasks, 25 mL transfer pipettes, 10 mL and 600 mL beakers, dropper, glass rods, tissue.
- $0.01 \text{ mol} \cdot \text{L}^{-1}$ hydrochloric acid (HCl) stock solution, benzoic acid, phenol, distilled water.

PROCEDURES

1. Analysis of standard solutions

Thoroughly clean out the UV-spectrometer cuvettes prior to analysis by rinsing them with acetone followed by deionized water. Use a large beaker to contain all the washings. Only handle the cuvettes on the edges, as touching the polished surfaces will leave grease and trace oils from your fingers that can interfere with the analyses.

Prepare a set of standard phenol solutions to cover the range of $5-20 \text{ mg} \cdot \text{L}^{-1}$. Weigh out accurately approximately 20 mg of phenol, making a note of the exact weight in mg to 1 decimal point (i. e. you will need to use a four figure balance to achieve this). Add this quantitatively to a 1 L volumetric and make up to the mark using $0.01 \text{ mol} \cdot \text{L}^{-1}$ hydrochloric acid (ensure all of the solid dissolves). Use this standard solution to prepare 50 mL of 5, 9, 12, 16, and 20 $\text{mg} \cdot \text{L}^{-1}$ phenol standard solutions by dilution, using $0.01 \text{ mol} \cdot \text{L}^{-1}$ hydrochloric acid as diluent, and appropriate volumetric flasks and pipettes.

Prepare a set of standard benzoic acid solutions to cover the range $2-10 \text{ mg} \cdot \text{L}^{-1}$

in the same way as for the phenol series, except that, this time you will weigh out accurately approximately 10 mg of the benzoic acid, again making a note of the exact weight in mg. Transfer this quantitatively to a 1 L volumetric flask. Make up to the mark with 0.01 mol • L^{-1} hydrochloric acid (ensure all of the solid dissolves). Use this standard solution to prepare 50 mL of 2, 4, 6, 8, and 10 mg • L^{-1} standard benzoic acid solutions by dilution with 0.01 mol • L^{-1} hydrochloric acid using appropriate volumetric flasks and pipettes.

The ten stock solutions (both phenol and benzoic acid) which you have now prepared cover a range of concentrations between 2—20 mg • L^{-1}. Transfer approximately 2 mL of each of the ten solutions to a small sample vial (label effectively), and place each one in the carrier provided for transporting them to the UV-spectrometer.

Ask the demonstrator in charge for instructions on the operating procedures for the UV-spectrometer.

First, make sure that there is a reference cell cuvette containing 0.01 M HCl already located in the UV-spectrometer (refer to the instruction board for this instrument). Next transfer the most dilutephenol sample to the quartz cuvette and place it into the sample cell holder. Set the instrument to scan over a range of excitation wavelengths from 200 nm to 500 nm. Repeat for the intermediate and higher concentrations, then plot all these UV spectra of the phenol standards on the same hardcopy.

Repeat the above exercise with the five benzoic acid samples in order to obtain all three UV spectra of the three benzoic acid samples on the same hardcopy.

Determine the absorbances at both 220 nm and 270 nm for all five phenol concentrations studied and record this data in your notebook. Use this tabulated absorbance data to produce a rough calibration plot of the intensity of UV absorbance vs. the concentration of each phenol solution at both wavelengths. The data should fit a straight line, according to the Beer-Lambert Law given in Equ. 1. If this is not the case, then you will need to repeat the preparation and recording of the UV spectrum from any sample which appears to be an "outlier". [Note: in addition to connecting together all five points from the standards, this straight line should also go through the origin].

Repeat the above exercise for the three UV spectra you have obtained from the benzoic acid standards.

For the final sample analysis appearing in your written report, all this data should be re-plotted using Excel and the line of best fit obtained for the four series of data: phenol at 220 nm; phenol at 270 nm; benzoic acid at 220 nm; and benzoic acid at 270 nm.

2. Analysis of a mixture of phenol and benzoic acid of unknown composition

The following part of the experiment should be performed in triplicate.

The demonstrator will provide you with a sample of a mixture of benzoic acid andphenol of unknown composition (labelled A, B, C or D). Record the UV-VIS spectrum of this unknown mixture and then determine the amount of phenol and benzoic acid in your sample by reference to the calibration graphs which you have produced in part 1. Since both phenol and benzoic acid show significant absorbances at both 220 nm and 270 nm, you will need to solve two simultaneous equations (i. e. equation 2 needs to be applied at each wavelength) when performing this calculation.

In addition, determine the reproducibility of the experiment for each component by calculating the range error 'E_r' using the following equation:

$$Er = \frac{(X_{highest} - X_{lowest})}{\overline{X}} \times 100 \qquad (2\text{-}25\text{-}3)$$

DATA TREATMENT

1. Analysis ofphenolstandard solutions

Flask	1	2	3	4	5
$c(\text{mg} \cdot \text{L}^{-1})$					
absorption A at 220 nm					
Equation $A_{220} \sim c$					
$\varepsilon_{220} (\text{L} \cdot \text{cm}^{-1} \cdot \text{mol}^{-1})$					
correlation coefficient r_{220}					
absorption A at 270 nm					
Equation $A_{270} \sim c$					
$\varepsilon_{270} (\text{L} \cdot \text{cm}^{-1} \cdot \text{mol}^{-1})$					
correlation coefficient r_{270}					

Analysis ofbenzoic acid standard solutions

Flask	1	2	3	4	5
$c(\text{mg} \cdot \text{L}^{-1})$					
absorption A at 220 nm					
Equation $A_{220} \sim c$					
$\varepsilon_{220} (\text{L} \cdot \text{cm}^{-1} \cdot \text{mol}^{-1})$					

(Continued)

Flask	1	2	3	4	5
correlation coefficient r_{220}					
absorption A at 270 nm					
Equation $A_{270}\sim c$					
ε_{270} ($L \cdot cm^{-1} \cdot mol^{-1}$)					
correlation coefficient r_{270}					

2. Drawing four standard curves

1. Analysis of unknown composition

Flask	A	B	C	D
absorption A at 220 nm				
absorption A at 270 nm				
c(phenol) ($mg \cdot L^{-1}$)				
c(benzoic acid) ($mg \cdot L^{-1}$)				

QUESTIONS

1. What are the units of ε, which you calculated in Part 1?

2. Why have you used a sample cell made of quartz in this experiment?

3. The K-band of benzene (Figure 2 – 30) has shifted to about 220 nm inphenol and benzoic acid. Suggest an explanation.

2.26 Identification of Benzoic Acid and Aniline by Infrared Spectroscopy

AIMS

1. To learn the sample preparation method for conventional samples.

2. To understand the principle of the infrared spectrometer.

INTRODUCTION

Infrared radiation (IR), sometimes called infrared light, is electromagnetic radiation (EMR) with longer wavelengths than those of visible light, and is therefore generally invisible to the human eye, although IR at wavelengths up to 1050 nanometers (nm)s from specially pulsed lasers can be seen by humans under certain conditions. IR wavelengths extend from the nominal red edge of the visible spectrum at 700 nanometers (frequency 430 THz), to 1 millimeter (300 GHz).

Usually IR is divided into three regions. One is the near-infrared region: the wavelength is 0. 75—2. 5 μm (wavenumber is 13 300—4 000 cm^{-1}). It consists of overtones and combination bands of the fundamental molecular absorptions found in the mid infrared region, and also is called the overtone region. Then, the mid-infrared region: the wavelength is 2. 5—50 μm (wavenumber is 4 000—200 cm^{-1}), also known as the vibration region. It is the spectroscopic fingerprint region of molecules since the energy of an mid-infrared photon corresponds to the fundamental ro-vibrational transitions within most molecules. Far-infrared region: the wavelength of 50—1 000 μm (wave number in the 200—10 cm^{-1}), also known as the rotating region. The mid-infrared region is the most studied and applied area.

Infrared spectroscopy involves the interaction of infrared radiation with matter. It covers a range of techniques, mostly based on absorption spectroscopy. As with all spectroscopic techniques, it can be used to identify and study chemical substances. The method or technique of infrared spectroscopy is conducted with an instrument called an infrared spectrometer (or spectrophotometer) to produce an infrared spectrum.

An IR spectrum can be visualized in a graph of infrared light absorbance (or transmittance) on the vertical axis vs. frequency or wavelength on the horizontal axis. In addition to the wavelength λ, the horizontal coordinates of the infrared spectrum is more commonly used as the wave number σ. The wave number is the reciprocal of the wavelength and indicates the number of waves contained in a unit of centimeter wavelength, with the symbol cm^{-1}. The relationship is:

$$\lambda = 1/\sigma \qquad\qquad (2\text{-}26\text{-}1)$$

The infrared spectrometer can be divided into dispersion type and interference type. The dispersive infrared spectrometer has prism and grating splitter, and the interference type is the Fourier transform infrared spectrometer (FTIR), shown in Figure 2 – 33. The most important difference is that there is no dispersion element in FTIR. FTIR is used in this experiment.

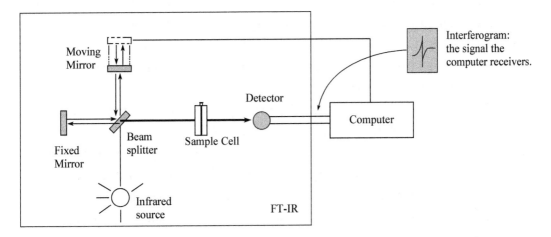

Figure 2 - 33   Schematic diagram of typical FTIR spectrophotometer.

FTIR spectroscopy is a measurement technique that allows one to record infrared spectra. Infrared light is guided through an interferometer and then through the sample (or vice versa). A moving mirror inside the apparatus alters the distribution of infrared light that passes through the interferometer. The signal directly recorded, called an "interferogram", represents light output as a function of mirror position. A data-processing technique called Fourier transform turns this raw data into the desired result (the sample's spectrum): Light output as a function of infrared wavelength (or equivalently, wavenumber).

Samples may be gas, liquid, or solid. Gaseous samples require a sample cell with a long pathlength to compensate for the diluteness. A simple glass tube with length of 5 to 10 cm equipped with infrared-transparent windows at the both ends of the tube can be used for concentrations down to several hundred ppm. Liquid samples can be sandwiched between two plates of a salt (commonly sodium chloride, or a common salt, although a number of other salts such as potassium bromide or calcium fluoride are also used). The plates are transparent to the infrared light and do not introduce any lines onto the spectra.

Solid samples can be prepared in a variety of ways. One common method is to crush the sample with an oily mulling agent (usually mineral oil Nujol). A thin film of the mull is applied onto salt plates and measured. The second method is to grind a quantity of the sample with a specially purified salt (usually potassium bromide) finely (to remove scattering effects from large crystals). This powder mixture is then pressed in a mechanical press to form a translucent pellet through which the beam of the spectrometer can pass. A third technique is the "cast film" technique, which is used mainly for polymeric materials. The sample is first dissolved in a suitable, non hygroscopic solvent. A drop of this solution is deposited on surface of KBr or NaCl cell. The solution is then evaporated to dryness and the film formed on the cell is analysed directly. Care is important to ensure that the film is not too thick otherwise light cannot pass through. The final method is to use a microtomy to cut a thin (20 – 100 μm) film from a solid sample.

The infrared spectrum of a sample is recorded by passing a beam of infrared light through the sample. When the frequency of the IR is the same as the vibrational frequency of a bond or collection of bonds, absorption occurs. Examination of the transmitted light reveals how much energy was absorbed at each frequency (or wavelength). The energies are affected by the shape of the molecular potential energy surfaces, the masses of the atoms, and the associated vibronic coupling. This measurement can be achieved by scanning the wavelength range using a monochromator. Alternatively, the entire wavelength range is measured using a Fourier transform instrument and then a transmittance or absorbance spectrum is generated using a dedicated procedure.

This technique is commonly used for analyzing samples with covalent bonds. Simple spectra are obtained from samples with few IR active bonds and high levels of purity. More complex molecular structures lead to more absorption bands and more complex spectra.

According to the relationship between infrared spectrum and molecular structure, each characteristic absorption band in the spectrum corresponds to the vibrational form of the particle or group of a certain compound. Therefore, the number, location, shape and intensity of the characteristic absorption band depend on the vibrational form and chemical environment of each group (chemical bond) in the molecule. As long as the vibration frequency (group frequency) and its displacement law of various groups are known, the relationship between the vibrational frequency and the molecular structure can be used to determine the attribution of the absorption band, and determine the group or bonds contained in the molecule, and then further estimate the molecular structure from the displacement of the characteristic vibrational frequency, the intensity and shape change of the spectral band.

Infrared spectroscopy is a simple and reliable technique widely used in both organic and inorganic chemistry, in research and industry. It is used in quality control, dynamic measurement, and monitoring applications. It is also used in forensic analysis in both criminal and civil cases.

In this experiment, you need to identify benzoic acid and aniline, one solid and one liquid sample. Pay attention to the different sample preparations and find out their characteristic vibrational peaks.

REAGENTS AND APPARATUS

• IR spectrometer, manual IR pellet tablet presser, agate mortar, microscale medicinal spoon, dropper.

• Benzoic acid, aniline, potassium bromide (spectral purity, KBr), ethanol, distilled water.

PROCEDURES

Turn on the infrared spectrometer and stabilize for about 5 minutes, and enter the corresponding IR workstation. Then, get a background spectrum.

Take 1 – 2 mg benzoic acid, add 100—200 mg potassium bromide powder, in agate mortar fully ground (about 2 μm), make it evenly mixed, and bake it under infrared light for about 10 min. About 80 mg mixture is taken out and spread evenly in a clean mode, and pressed under a pressure of 12—15 MPa for 1 minute to make a transparent pellet. The pellet is mounted on a solid sample holder, and the sample holder is then inserted into the sample cell of the infrared spectrometer. The wavenumber is scanned from 4 000—400 cm^{-1} to obtain the absorption spectrum for benzoic acid. Save the data.

Make a blank pellet, and then add 1 drop of aniline solution. The other operation steps are the same as mentioned above to obtain the infrared absorption spectrum of aniline.

End the experiment and close the workstation, infrared spectrometer and computer sequentially. Clean the IR lab.

DATA TREATMENT

1. The IR spectra of benzoic acid and aniline;

2. Data analysis by assigning the observed absorption frequency bands in the sample spectrum to appropriate normal modes of vibrations in the molecules.

QUESTIONS

1. When using the tableting method, why do you need to grind the solid sample to a particle size of about 2 μm?

2. Why do you want KBr powder to be dry to avoid water absorption and moisture?

2.27 Determination of Calcium Content in Water Using Atomic Absorption Spectrometry

AIMS

1. To understand the technique of atomic absorption spectrometry and be aware of the range of everyday uses of a modern atomic absorption spectrometer.

2. To know how to use standard addition method to calibrate.

INTRODUCTION

There is a number of methods for analyzing calcium in water. A traditional method in the quality control of calcium is complexometric titration. The concentration of calcium in natural waters is determined by spectrophotometric method, indirect potentiometric method, single-use optical sensor, capillary electrophoresis, and ion chromatography. However, atomic absorption spectrometry (AAS) is one of the most extensively used techniques for the determination of various elements with significant precision and accuracy. This analytical technique is remarkable for its selectivity, speed and fairly low operational cost.

The arrangement of the electrons within an atom is a unique feature of the structure of each element. Electrons may be excited from their normal ground state (valency) positions by the acquisition of energy supplied by an electrical discharge, heat from a flame, or by the absorption of electromagnetic radiation (light) of the appropriate wavelength. In the technique of atomic absorption spectrophotometry, atoms in their ground state are irradiated with light appropriate for the excitation of their electrons.

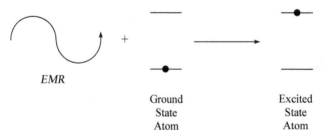

Figure 2 – 34 The Atomic Absorption Process

A population of ground state atoms is produced by aspirating a solution containing the material to be analysed, into a flame. An air-acetylene flame burns at 2 200 K which is sufficient to dissociate compounds into atoms, but does not produce a high proportion of excited atoms directly from the energy of the flame. Various devices are used to render the geometry and combustion characteristics of the flame stable, and so it is possible to direct a beam of radiation through the flame and to monitor the extent of absorption by the atom population present.

Figure 2 – 35   Close-up of the burner of the spectrometer to be used in the experiment. The acetylene cylinder is for safety reasons stored outside the building.

The radiation which is directed through the flame must be specific for the excitation of the atoms which are present in the sample to be analysed. If the sample is being analysed for the presence of zinc, then the incident radiation must contain wavelengths which are appropriate for the excitation of electrons within the zinc atom. The source of radiation is known as a hollow cathode lamp and is constructed from an evacuated glass envelope filled with an inert gas (typically neon), with a cathode made from the element to be analysed. When the lamp is operating it produces an emission consisting of the emission spectra of both the fill gas and of the element forming the cathode. Multi-element lamps may be made by forming the cathode from an alloy to obtain the spectral emission from all the elements within the alloy-lamps of this type are, however, usually less intense than those employing single element cathodes.

The total emission from a hollow cathode lamp therefore contains a mixture of wavelengths and it is usual to select a specific wavelength at which to perform the analysis. A monochromator is therefore used to isolate a suitable wavelength and the intensity of the radiation transmitted by the atoms in the flame is recorded by a photomultiplier detector.

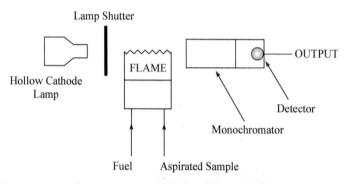

Figure 2 − 36   Basic Layout of an Atomic Absorption Spectrophotometer

Figure 2 − 37   The set of lamps available in the spectrometer you will use.
The appropriate lamp for the element to be analysed should be selected.

The relationship between the transmitted light intensity and the concentration of atoms in the flame is given by the Beer-Lambert law:

$$A = \log \frac{I_o}{I_t} = \varepsilon_\lambda c l \qquad (2\text{-}27\text{-}1)$$

Where: A is absorbance, I_o is the incident light intensity at wavelength λ, I_t is the transmitted light intensity at wavelength λ, ε_λ is the absorptivity at wavelength λ, c is concentration of ground state atoms in the flame, l is the optical path length of the flame.

The relationship between the absorbance and the concentration of the atoms in the flame should be linear. However, in practice deviations are found at high and low concentrations. This means that a preliminary calibration of absorbance vs. concentration is necessary. A calibration plot is constructed by measuring the absorbances of a series of solutions containing known amounts of the element to be analysed. The unknown sample is then diluted by trial and error and the measured absorbance inserted onto the calibration graph. The concentration of the original sample may then be determined after applying the dilution factors in reverse.

REAGENTS AND APPARATUS

• Atomic absorption spectrophotometer, calciumhollow cathode lamp, 50 mL and 500 mL volumetric flasks, 5.0 mL pipettes, droppers.

• water sample, calcium carbonate ($CaCO_3$), 6 mol • L^{-1} hydrochloric acid (HCl), distilled water.

PROCEDURES

1. Calcium stock solution preparation

Weigh accurately (use an analytical balance) about 2.5000 g of calcium carbonate, and cautiously add, dropwise, 9.00 mL 6 mol • L^{-1} hydrochloric acid. After the calcium carbonate is completely dissolved, then transfer the solution to a 1 000 mL volumetric flask (flask A) and dilute to volume with distilled water. Transfer 10.00 mL of this solution from flask A to a 1 000 mL volumetric flask (flask B), add 10.00 mL 6 mol • L^{-1} hydrochloric acid, dilute to volume with distilled water and mix well.

2. "Five" analyte solution for standard addition method

Transfer 10.00 ml of water sample to five 50 mL volumetric flask, then add 0; 2.00, 4.00, 6.00 and 8.00 calcium stock solution from flask B, respectively, dilute to volume with distilled water.

3. Calcium determination

Open the atomic absorption spectrophotometer, and open the corresponding gas. Choose the calcium hollow cathode lamp, and select the correct wave length. Set up other instrumental parameters and ignite. Determine the calcium content with the five calcium analyte solutions from low to high concentration.

The samples will be measured as a group. Calculate the calcium content of the water sample expressing your answer as mg L^{-1} (ppm; mg mL^{-1}).

Using the analytical results obtained by a typical class (which can be obtained from the technician running the samples) calculate the mean and standard deviation from the following formulae.

For n analyses: x_1, x_2, $\cdots$, x_i, $\cdots$, x_n

$$Mean = \bar{x} = \frac{\sum x_i}{n} \qquad (2\text{-}27\text{-}2)$$

$$Standard\ deviation = \sigma = \sqrt{\frac{\sum (x_i^2) - (n * (\bar{x})^2)}{(n-1)}} \qquad (2\text{-}27\text{-}3)$$

DATA TREATMENT

1. Data of calcium standard solutions

	concentration (mg L^{-1})	absorbance 1 (a. u.)	absorbance 2 (a. u.)	absorbance 3 (a. u.)	average of absorbance
Addition Solu Ca$_1$					
Addition Solu Ca$_2$					
Addition Solu Ca$_3$					
Addition Solu Ca$_4$					
Addition Solu Ca$_5$					

2. Calibrate theBeer-Lambert law
3. Data ofcalcium content

calcium content (mg L^{-1})	
data from other groups (mg L^{-1})	
mean $\bar{c}$ (mg L^{-1})	
standard deviation	

QUESTIONS

1. Can the atomic absorption spectrum be used for multicomponent measurement? Explain.

2. What other methods are used for calibration curve?

3. Compare your result with the mean and standard deviation for the given class data; comment on the validity of your result.

2.28 Fluorescence Determination of The Content of Hydroxybenzoic Acid Isomers

AIMS

1. To learn the basic principles and operation of fluorescence analysis.

2. To master how to determine the multi-component content using fluorescence.

INTRODUCTION

Organic compound molecules at the ground-state absorb the quantized energy, electrons will jump from the lower molecular orbital level to the higher orbital energy level, that is from the ground state into the excited states. The molecules at the excited state can release energy in the form of light radiation, in this process there will be light emitting phenomenon, which is called fluorescence or luminescence phenomenon.

Excitation: $S_0 + h\nu_{ex} \longrightarrow S_1$

Fluorescence (emission): $S_1 \longrightarrow S_0 + h\nu_{ex} + \text{heat}$

Here $h\nu$ is a generic term for photon energy with h is Planck's constant and ν is frequency of light. The specific frequencies of excited and emitted lights are dependent on the particular system. S_0 is called the ground state of the fluorophore (fluorescent molecule), and S_1 is its first (electronically) excited singlet state.

A molecule in S_1 can relax by various competing pathways, shown in Figure 2 – 38. It can undergo non-radiative relaxation in which the excitation energy is dissipated as heat (vibrations) to the solvent, or some other routes such as external conversion, internal transformation, intersystem crossing, vibration relaxation. Excited organic molecules can also relax via conversion to a triplet state, which may subsequently relax via phosphorescence, or by a secondary non-radiative relaxation step. Relaxation from S_1 can also occur through interaction with a second molecule through fluorescence quenching.

Fluorescence is the emission of light by a substance that has absorbed light or other electromagnetic radiation. It is a form of luminescence. In most cases, the emitted light has a longer wavelength, and therefore lower energy, than the absorbed radiation. The most striking example of fluorescence occurs when the absorbed radiation is in the ultraviolet region of the spectrum, and thus invisible to the human eye, while the emitted light is in the visible region, which gives the fluorescent substance a distinct color that can be seen only when exposed to UV light. Fluorescent materials cease to glow nearly immediately when the radiation source stops, unlike phosphorescent materials, which continue to emit light for some time after.

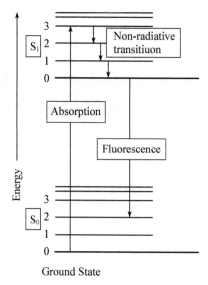

Figure 2 – 38　Fluorescence Spectroscopy

Fluorescence has many practical applications, including mineralogy, gemology, medicine, chemical sensors (fluorescence spectroscopy), fluorescent labelling, dyes, biological detectors, cosmic-ray detection, and, most commonly, fluorescent lamps. Fluorescence also occurs frequently in nature in some minerals and in various biological states in many branches of the animal kingdom.

Fluorescence spectroscopy (also known as fluorimetry or spectrofluorometry) is a type of electromagnetic spectroscopy that analyzes fluorescence from a sample. It involves using a beam of light, usually ultraviolet light, that excites the electrons in molecules of certain compounds and causes them to emit light; typically, but not necessarily, visible light. A complementary technique is absorption spectroscopy. In the special case of single molecule fluorescence spectroscopy, intensity fluctuations from the emitted light are measured from either single fluorophores, or pairs of fluorophores.

Fluorescence spectroscopy uses such scheme (shown in Figure 2 – 39): The light from an excitation source passes through a filter or monochromator, and strikes the sample. A proportion of the incident light is absorbed by the sample, and some of the molecules in the sample fluoresce. The fluorescent light is emitted in all directions. Some of this fluorescent light passes through a second filter or monochromator and reaches a detector, which is usually placed at 90° to the incident light beam to minimize the risk of transmitted or reflected incident light reaching the detector. This feature adds the high sensitivity of fluorescence applications is the spectral selectivity.

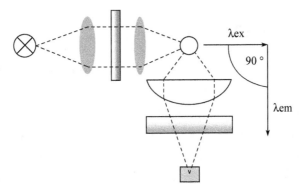

Figure 2 - 39   A simple design of the components of a fluorescence spectroscopy

At low concentrations the fluorescence intensity will generally be proportional to the concentration of the fluorophore.

Ortho-hydroxybenzoic acid (also known as salicylic acid) and the meta-hydroxybenzoic acid have similar molecular components, all containing a benzene ring which has fluorescence, but fluorescence properties is different due to their different position of the substituents (Figure 2 - 40).

In pH=12 alkaline solution, they both have fluorescence under the excitation of ultraviolet light near 310 nm. In near neutral pH=5. 5 solution, m-hydroxybenzoic acid does not have fluorescence, but o-hydroxybenzoic acid has the same intensified fluorescence due to the intermolecular hydrogen bonding. Using this property, the concentration of o-hydroxybenzoic acid in the mixture can be determined at pH=5. 5, with the presence of m-hydroxybenzoic acid. Fluorescence intensity at pH = 12 is further measured with the same amount of mixture solution. By subtracting the fluorescence intensity of o-hydroxybenzoic acid measured at pH = 5. 5, then the concentration of m-hydroxybenzoic acid can obtained as well.

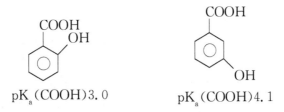

pK_a(COOH)3. 0 pK_a(COOH)4. 1

Figure 2 - 40   Structure of ortho-hydroxybenzoic acid and meta-hydroxybenzoic acid

In this experiment, you will first do the calibration curve of ortho-hydroxybenzoic acid and meta-hydroxybenzoic acid, respectively. Then, determine the unknown hydroxybenzoic acid mixture both at pH = 5. 5 and pH 12. Their concentrations will be obtained from their respect calibration curve.

REAGENTS AND APPARATUS

- Fluorescence spectrophotometer, fluorescence cuvette, 50 mL volumetric flasks, 5.0 mL and 25 mL measuring pipettes, droppers.
- 0.1 mol • L^{-1} sodium hydroxide (NaOH) solution, 0.1 mol • L^{-1} HAC-NaAC buffer solution (pH = 5.5), 60 μg • mL^{-1} o-hydroxybenzoic acid stock aqueous solution in 0.1 mol • L^{-1} HAC-NaAC solution, 60 μg • mL^{-1} m-hydroxybenzoic acid stock aqueous solution in 0.1 mol • L^{-1} NaOH buffer solution, respective unknown mixture sample in 0.1 mol • L^{-1} NaOH solution and 0.1 mol • L^{-1} HAC-NaAC buffer solution, distilled water.

PROCEDURES

1. Prepare a set of o-hydroxybenzoic acid and m-hydroxybenzoic acid standard solutions

Use o-hydroxybenzoic acid stock solution to prepare 50 mL of 5, 10, 15, 20, and 30 μg • mL^{-1} o-hydroxybenzoic acid standard solutions by dilution, using 0.1 mol • L^{-1} sodium hydroxide (NaOH) solution as diluent, and appropriate volumetric flasks and pipettes.

Use m-hydroxybenzoic acid stock solution to prepare 50 mL of 5, 10, 15, 20, and 30 μg • mL^{-1} m-hydroxybenzoic acid standard solutions by dilution, using 0.1 mol • L^{-1} HAC-NaAC buffer solution as diluent, and appropriate volumetric flasks and pipettes.

2. Analysis of standard solutions

Thoroughly clean out the fluorescence cuvettes prior to analysis by rinsing them with acetone followed by deionized water. Use a large beaker to contain all the washings. Only handle the cuvettes on the edges, as touching the polished surfaces will leave grease and trace oils from your fingers that can interfere with the analyses.

Open the fluorescence devices, and corresponding software in the computer. Transfer the most dilute o-hydroxybenzoic acid standsolution to the cuvette and place it into the sample cell holder. Set a fixed excitation wavelength, and scan for a range of emission wavelengths. From the emission spectra, find the maximum emission wavelength, and then scan for the closest matched excitation wavelength. Repeated several times, the accurate maximum excitation wavelength is obtained.

Then set the instrument with the accurate excitation wavelength, and scan over a range of emission wavelengths. Repeat for the other higher concentrations, then plot all these fluorescence spectra of the o-hydroxybenzoic acid standards on the same hardcopy.

Repeat above steps for the analysis of m-hydroxybenzoic acid standard solutions.

3. Analysis of the mixed sample

Thoroughly clean out the fluorescence cuvettes again. Record the fluorescence spectra of the unknown mixture in different buffer and then determine the amount of o-hydroxybenzoic and m-hydroxybenzoic acid in your sample by reference to the calibration graphs which you have produced. The amount of o-hydroxybenzoic acid can be directly obtained from the spectrum from unknown sample measured at pH= 5.5 buffer and o-hydroxybenzoic acid calibration plot. By subtracting the fluorescence intensity of o-hydroxybenzoic acid measured at pH=5.5, then the concentration of m-hydroxybenzoic acid can obtained from its calibration plot as well.

DATA TREATMENT

1. Analysis of standard solutions

	concentration $\mu g \cdot mL^{-1}$	intensity a. u.		concentration $\mu g \cdot mL^{-1}$	intensity a. u.
o-hydroxy benzoic acid	5		m-hydroxy benzoic acid	5	
	10			10	
	15			15	
	20			20	
	30			30	

2. Draw the calibration plots
3. Analysis of the mixed sample

Intensity at pH 5.5 (a. u.)	
Intensity at pH 12 (a. u.)	
o-hydroxybenzoic acid concentration ($\mu g \cdot mL^{-1}$)	
m-hydroxybenzoic acid concentration ($\mu g \cdot mL^{-1}$)	

QUESTIONS

1. What do λ_{ex}^{max} and λ_{em}^{max} represent? Why should the λ_{ex}^{max} and λ_{em}^{max} fluorescence intensities be essentially the same for a component?

2. What is the main acid and alkali form of o-hydroxybenzoic acid and m-hydroxybenzoic acid in aqueous solution at pH 5.5? Why are the fluorescence properties of the two different?

3. Summarize several factors that affect the fluorescence intensity of the substance from this experiment.

2.29 Matrix-Assisted Laser Desorption/Ionization (MALDI) Mass Spectrometry (MS) Analysis of Oligosaccharide

AIMS

1. To know how to prepare a biomolecular sample for analysis by MALDI mass spectrometry.

2. To know how to acquire MALDI mass spectra.

3. To be able to extract mass information from a MALDI mass spectrum of a simple "unknown" (bio)polymer.

4. To be able to calculate and report mass resolution, and accuracy and precision of a mass measurement.

INTRODUCTION

Mass spectrometry (MS) is an analytical laboratory technique to separate the components of a sample by their mass and electrical charge. The instrument used in MS is called mass spectrometer. It produces a mass spectrum that plots the mass-to-charge (m/z) ratio of compounds in a mixture. MS is used for both qualitative and quantitative chemical analysis. It may be used to identify the elements and isotopes of a sample, to determine the masses of molecules, and as a tool to help identify chemical structures. It can measure sample purity and molar mass.

In MS, gas-phase molecular analyte ions have to be generated as these are the molecular species that can be manipulated (e. g. accelerated and/or diverted) and thus separated and analysed by magnetic and electric fields. Thus, a mass spectrometer is generally made up of three parts: (1) the ion source, (2) the mass analyser and (3) the detector (see Figure 2 – 41).

The initial sample may be a solid, liquid, or gas. The sample is vaporized into a gas and then ionized by the ion source, usually by losing an electron to become a cation. Even species that normally form anions or don't usually form ions are converted to cations (e. g. , halogens like chlorine and noble gases like argon). The ionization chamber is kept in a vacuum so the ions that are produced can progress through the instrument without running into molecules from air. Ionization is from electrons that are produced by heating up a metal coil until it releases electrons. These electrons collide with sample molecules, knocking off one or more electrons. Since it takes more energy to remove more than one electron, most cations produced in the ionization chamber carry a $+1$ charge. A positive-charged metal plate pushes the sample ions to the next part of the machine.

There are many ionization techniques available such as electron ionization (EI), chemical ionization (CI), and electrospray ionization (ESI), and the best choice of these depends on the type of analyte to be analysed. The two major ionization techniques in biological MS are ESI and matrix-assisted laser desorption/ionization (MALDI), which are soft ionization techniques. In this ionization method samples are fixed in a crystalline matrix and are bombarded by a laser. The sample molecules vaporize into the vacuum while being ionized at the same time without fragmenting or decomposing (see Figure 2 – 42).

Once molecular ions are produced they are transferred to the mass analyser. In the mass analyzer, the ions are then accelerated through a potential difference and focused into a beam. The purpose of acceleration is to give all species the same kinetic energy, like starting a race with all runners on the same line. The ion beam passes through a magnetic field which bends the charged stream. Lighter components or components with more ionic charge will deflect in the field more than heavier or less charged components. Here are many types of mass analysers used for the essential ion manipulations such as quadrupoles, ion traps, orbitraps and time-of-flight mass analysers. All of these enable the separation and measurement of the ion's molecular mass by exploiting the ion's mass-to-charge (m/z) ratio. For example, time-of-flight (TOF) mass analyser separates ions by their m/z ratio and determines that m/z ratio by the time it takes for the ions to reach a detector.

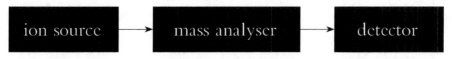

Figure 2 – 41　The essential parts and workflow of a mass spectrometer.

A detector counts the number of ions at different deflections. The data is plotted as a graph or spectrum of different masses. Detectors work by recording the induced charge or current caused by an ion striking a surface or passing by. Because the signal is very small, an electron multiplier, Faraday cup, or ion-to-photon detector may be used. The signal is greatly amplified to produce a spectrum.

In short, a mass spectrometer measures the molecular ion's m/z ratio. Knowing the ion's charge state its mass can be determined.

In this practical, you will prepare oligosaccharide samples for MALDI MS analysis by using a standard MALDI target plate, on which the MALDI samples will be spotted. You will have the opportunity to learn more about the practical details and challenges in modern mass spectrometry. The analysis of the obtained mass spectra will enable you to understand the basic features of a biomolecular mass spectrum and its interpretation. From replicate measurements you will determine the mass measurement accuracy and precision.

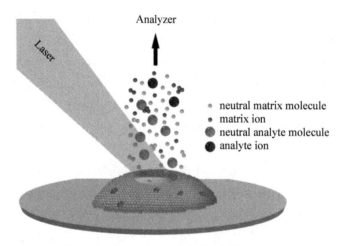

Figure 2 – 42   Scheme of the MALDI process.

REAGENTS AND APPARATUS

• MALDI-TOF or MALDI-Q-TOF mass spectrometer with MALDI target plates, balance, small-volume pipettes, vortexer, wash bottles, microcentrifuge tubes, pipette tips, lint-free tissues.

• Analytes, 20 mg • mL^{-1} 2, 5 – dihydroxybenzoic acid (DHB) solution, 10 mg • mL^{-1} alpha-cyano – 4 – hydrocycinnamic acid (CHCA) in 50% acetonitrile/ 50% water, 30% hydrogen peroxide solution (H$_2$O$_2$) solution, acetonitrile, ethanol, distilled water.

PROCEDURES

1. Introduction to the experiment, MALDI MS, and risk assessments

A short introduction to the experiment and its activity schedule will be given. MALDI MS will be introduced and discussed with regard to the instrument and analyte used for the experiment. Risk assessments will be completed and signed off.

2. Preparation of MALDI samples

At the start of the practical session, the demonstrator will demonstrate the preparation of samples for MALDI MS analysis. You will repeat the MALDI sample preparation in pairs, preparing four replicate MALDI samples for each set of samples.

There are three sets of four samples to be prepared:

One MALDI samples set of analyte solution mixed with CHCA matrix solution;

One MALDI samples set of analyte solution mixed with DHB matrix solution;

One MALDI samples set of analyte solution mixed with DHB matrix solution and treated with 0.5% H_2O_2.

For the matrix solutions, you prepare two solutions: an aqueous solution (using HPLC-grade water) of 2,5 - dihydroxybenzoic acid (DHB) at a concentration of approximately 20 mg • mL^{-1} and an alpha-cyano - 4 - hydrocycinnamic acid (CHCA) solution in 50% acetonitrile/50% water (all HPLC-grade or higher purity) at a concentration of approximately 10 mg • mL^{-1} by weighing out the matrix compounds and adding the solvents using 1.5/2 mL microcentrifuge tubes.

In addition, you have to prepare an aqueous 0.5% H_2O_2 solution ($\sim$ 50—100 µL) using a 30% H_2O_2 stock solution. Warning: The steps of dissolving matrix compounds in acetonitrile and preparing the 0.5% H_2O_2 solution have to be undertaken in a fume cabinet using gloves.

The analyte solution for all MALDI samples is a diluted stock solution of a standard oligosaccharide mixture, containing various oligosaccharides in the mass range of 300—2 000 Da at micromolar concentration. The analyte solution will be provided by the demonstrator. Warning: In general, you should treat any 'unknown' analyte as highly toxic because you do not know what it is. For this practical and the risk assessment/COSHH form, the type of 'unknown' analyte solution is a standard oligosaccharide sample at low concentration.

Before preparing the MALDI samples, you need to clean the MALDI target plates with alternating washes of pure water and ethanol from wash bottles and wiping/drying the plate with lint-free tissue using gloves. A final wash should be undertaken with water but the air dried rather than using wipes.

To prepare the MALDI samples, a volume of 0. 5 μL of matrix solution is spotted for each sample on the clean MALDI target plate. You can spot an entire set of four matrix solution droplets with one tip before changing it. Next you add a volume of 0. 5 μL of the analyte solution on top of each matrix solution droplet but you have to change the tip between samples to avoid cross-contamination of your solutions.

After all matrix solutions are spotted and mixed with the analyte solutions you leave them dust-protected to dry.

One of the three sets of samples (see above) will then be treated with 0. 5% H_2O_2 by spotting a volume of 0. 5 μL of the 0. 5% H_2O_2 solution on top of the dry MALDI samples. The target plate with the re-dissolved MALDI samples will be left again to dry and once the samples are dry again all can be analysed by MALDI MS.

3. Demonstration of MS instrument calibration and the acquisition of MALDI mass spectra

The demonstrator will explain the operation of the various components of the MALDI mass spectrometer and perform an instrument calibration using a standard peptide mixture. All students will then acquire in pairs their own mass spectra of their 12 samples by summing up 500 single-shot spectra obtained from 100 laser shots each at 5 different desorption positions for each sample. The monoisotopic ion signal intensity of the protonated peptide ion species need to be recorded for all 12 sum spectra. Hardcopies or e-copies of the spectral data will be made available for report writing.

4. Workshop on the mass spectral data analysis and discussion of MS experiments

In this workshop the acquired MS data including some of its spectral features (e. g. analyte and matrix ion signals, noise, resolution) will be discussed. General interpretation of MALDI and ESI spectra will be provided and applied to other spectra acquired by these and other ionization techniques. Mass measurement accuracy and precision will be explained and the students will use their own data to determine the mass measurement accuracy and precision, using descriptive measurement statistics (i. e. mean $\bar{x}$ and variance σ^2):

$$\bar{x} = \frac{1}{N} \sum_{i=1}^{N} x_i \tag{2-29-1}$$

$$\sigma^2 = \frac{1}{N} \sum_{i=1}^{N} (x_i - \bar{x})^2 \tag{2-29-2}$$

DATA TREATMENT

1. The MS spectra of three sets of four samples.
2. MS spectral data analysis.
3. Calculate the mass measurement accuracy and precision:

	sample 1	sample 2	sample 3	sample 4		2
matrix 1						
matrix 2						
matrix 3						

QUESTIONS

1. What is the amount of analyte in one miroliter if the analyte concentration is 10 micromolar?

2. What is typically the main reason for the isotopic pattern of organic compounds in mass spectra?

3. What is the 'unit' used to describe the experimental values in MS measurements and what 'unit' can be used to describe molecular mass in general?

4. Why do lighter ions reach the detector in a TOF mass spectrometer earlier than heavier ions?

2. 30   Analysis Methods of Nuclear Magnetic Resonance (NMR) Spectroscopy

AIMS

1. To know how to prepare a sample of an organic compound for analysis by ^{1}H and ^{13}C NMR spectroscopy.

2. To know how to acquire both a ^{1}H and a ^{13}C NMR spectrum.

3. To be able to interpret both the ^{1}H and ^{13}C NMR spectrum of a simple "unknown" organic compound in conjunction with its IR and/or MS spectra.

INTRODUCTION

Nuclear magnetic resonance (NMR) is a physical observation in which nuclei in a strong constant magnetic field are perturbed by a weak oscillating magnetic field and respond by producing an electromagnetic signal with a frequency characteristic of the magnetic field at the nucleus. This process occurs near resonance, when the oscillation frequency matches the intrinsic frequency of the nuclei, which depends on the strength of the static magnetic field, the chemical environment, and the magnetic properties of the isotope involved.

Nuclear magnetic resonance spectroscopy, most commonly known as NMR spectroscopy, is a spectroscopic technique to observe local magnetic fields around atomic nuclei. It is a spectroscopy technique which is based on the absorption of electromagnetic radiation in the radio frequency region 4 to 900 MHz by nuclei of the atoms. Over the past fifty years, NMR has become one of the techniques most commonly employed by chemists for determining the structure of organic compounds. Of all the spectroscopic methods, it is the only one for which a complete analysis and interpretation of the entire spectrum is normally expected.

NMR spectrometer usually contains seven parts, as shown in Figure 2 – 43. Sample holder, is a glass tube with 8. 5 cm long, 0. 3 cm in diameter. Permanent magnet provides homogeneous magnetic field at 60—100 MHZ. Magnetic coils induce magnetic field when a current flows through them. Sweep generator is to produce the equal amount of magnetic field pass through the sample. Radio frequency transmitter coil produces a short powerful pulse of radio waves. Radio frequency receiver coil that detects radio frequencies emitted as nuclei relax to a lower energy level. Read out system usually is a computer that analyses and records the data.

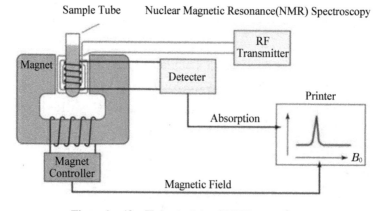

Figure 2 – 43   The principle of NMR spectroscopy

NMR spectroscopy can provide detailed and quantitative information on the functional groups, topology, dynamics and three-dimensional structure of molecules in solution and the solid state. NMR spectroscopy can obtain physical, chemical, electronic and structural information about molecules due to the chemical shift of the resonance frequencies of the nuclear spins in the sample. Peak splittings due to J-or dipolar couplings between nuclei are also useful. Since the area under an NMR peak is usually proportional to the number of spins involved, peak integrals can be used to determine composition quantitatively.

In this practical, you will use both ^{1}H and ^{13}C NMR spectra in order to determine the structures of 'unknown' compounds.

A summary for NMR is given below:

The peaks appearing in a ^{1}H NMR spectrum can be analysed for the following three kinds of information:

a) *Chemical shift* (0—12 ppm). This gives an indication of the kind of functional groups in the immediate chemical environment of a proton (e. g. aliphatics resonate at 1—2 ppm; 3—4 ppm for substitution by electronegative atoms such as O, N or Cl; 5—7 ppm for alkenes; 6—8 ppm for aromatics; and 8—12 ppm for carboxylic acids);

b) *Splitting pattern*. This gives an indication of how many protons are in the immediate vicinity of the proton being observed. If n protons are "close by" (normally this means connected by 3 bonds), then the peak will be split into an $n+1$ multiplet;

c) *Intensity*. The area under each ^{1}H NMR peak is usually displayed as an integral, indicating the relative number of protons in a given chemical environment.

^{13}C NMR spectra normally provide only chemical shift information, but over a wider range: 0—220 ppm (there is no splitting and signal intensity is not a very reliable guide to the number of carbons present in a given environment-however, note that quaternary carbons can often be recognized by their low intensity).

REAGENTS AND APPARATUS

- NMR Spectrometer, pasteur pipettes, cotton wool filter, and vials.
- Unknown sample, deuteriated solvent.

PROCEDURES

1. Preparation of an "unknown"NMR sample

At the start of the practical session, the demonstrator will demonstrate the preparation of an "unknown" compound in an appropriate deuteriated solvent, ready for both ^{1}H and ^{13}C NMR spectroscopy.

2. Demonstration of the acquisition of NMR spectra

The demonstrator will explain the operation of the various components of the NMR spectrometer (superconducting magnet, probe, preamplifier, acquisition console and acquisition computer). Students should also consult the handout for a summary of this description.

The demonstrator will demonstrate the procedure for the acquisition of a^{1}H NMR spectrum (loading the sample in the magnet, locking, shimming, setting the receiver gain, loading the pulse programme, Fourier transform, phasing, referencing, integration, expansion and plotting) for this same sample. A ^{13}C NMR spectrum of the "unknown" will also then be acquired. Students should consult the handout for a summary of this process, when answering point 2 in the assessment.

3. Worked examples of the interpretation of NMR spectra

The interpretation of both the ^{1}H NMR spectrum and the ^{13}C NMR spectrum of the 'unknown' compound, which was acquired in the previous section will be explained by the demonstrator. Several other examples of the interpretation of NMR spectra will be given, some in conjunction with IR and MS spectra.

4. Workshop on the interpretation of NMR spectra

Students will be provided with a set of spectroscopic problems, similar to the worked examples discussed by the demonstrator in the previous section, which they are expected to solve working either individually or in small groups in a workshop environment, at which the demonstrator will be available for consultation. At the end of the workshop, a set of fully-worked model answers for each problem will be distributed. Students will also receive a set of spectra for an "unknown" organic compound which will form part of the basis of the assessment of the practical.

DATA TREATMENT

1. The ^{1}H NMR and ^{13}C NMR spectra of the unknown sample.

2. ^1H NMR and ^{13}C NMR spectral data analysis.

QUESTIONS

1. What are the necessary conditions for producing NMR?

2. What factors will cause errors in the experiment?

3. Briefly describe the causes of coupling and spin splitting. What does the magnitude of coupling represent, and How to measure?

4. What information does NMR spectroscopy provide for structural analysis of organic compounds?

2.31 Potentiometric Titration of Phosphoric Acid

AIMS

1. To learn the method of potentiometric titration and determining stoichiometric points.

2. To determine of the concentration of phosphoric acid.

INTRODUCTION

Potentiometric titrations are a useful method of determining unknown concentrations in many different types of chemical systems. They may be employed in precipitation titrations, complex formation titrations, acid-base titrations and oxidation/reduction titrations. Potentiometric titrations provide more reliable data than data from titrations that use chemical indicators and are particularly useful with colored or turbid solutions and for detecting the presence of unsuspected species.

Potentiometric titrations involve the measurement of the potential of a suitable indicator electrode with respect to a reference electrode as a function of titrant volume. Typical cells for measuring pH consist of a glass indicator electrode and a saturated calomel reference electrode immersed in the solution whose pH is unknown. The indicator electrode consists of a thin, pH sensitive glass membrane sealed onto one end of a heavy-walled glass or plastic tube. A small volume of hydrochloric acid saturated with silver chloride is contained in the tube. A silver wire in this solution forms a silver/silver chloride inner-reference electrode, which is connected to one of the terminals of the potential-measuring device, pH-meter. The calomel electrode is connected to the other terminal.

The electric potential created between the glass electrode, and the inner-reference electrode is a function of the pH value (activity of hydronium ion, a H_3O^+) of the measured solution. So once the potential difference between glass electrode and outer-reference calomel electrode has been measured the pH value can be calculated.

Modern pH electrodes are usually of the "combination" type, meaning that a single cylinder contains both a glass membrane electrode and the outer-reference calomel electrode, as shown in Figure 2 – 44.

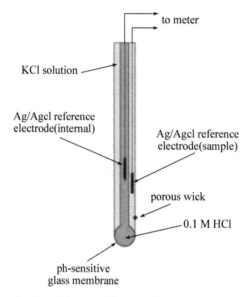

to meter

KCl solution

Ag/Agcl reference electrode(internal)

Ag/Agcl reference electrode(sample)

porous wick

0.1 M HCl

ph-sensitive glass membrane

Figure 2 – 44   Schematic diagram of pH combined electrode

A change in hydronium ion concentration causes a change in composition of the glass membrane due to an ion exchange process involving the solution and the membrane. A corresponding change in membrane potential, proportional to pH, is what is measured. All other potentials are constant. In effect the membrane potential (variable) is measured against two fixed potentials, the external reference and the internal reference, both Ag/AgCl reference electrodes. Potential difference is measured using a high impedance (internal resistance) potentiometer.

A typical set up for potentiometric titrations is given in Figure 2 – 45. Titration involves measuring and recording the cell potential (in units of millivolts or pH) after each addition of titrant. The titrant is added in large increments at the outset and in smaller and smaller increments as the end point is approached (as indicated by larger changes in response per unit volume). Sufficient time must be allowed for the attainment of equilibrium after each addition of the reagent by continuous stirring. For this a magnetic stirrer with a stirring magnet bar is used.

Phosphoric acid, H_3PO_4, is a polyprotic ("many hydrogen") acid. Although often listed together with strong mineral acids (hydrochloric, nitric and sulfuric), it is relatively weak, with $pK_{a1} = 2.15$, $pK_{a2} = 7.20$ and $pK_{a3} = 12.35$. That means titration curve contains only two inflection points and phosphoric acid can be titrated either as a monoprotic acid or as a diprotic acid. In the first case acid has to be titrated around pH 4.7 (for example methyl orange), in the second case around pH 9.6.

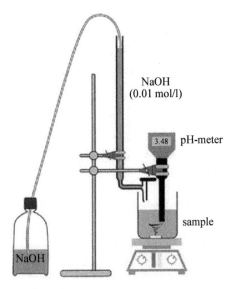

Figure 2 – 45   Setup of the potentiometric titration experiment.

In this experiment, you will conduct a potentiometric titration of a solution of phosphoric acid (H_3PO_4) of unknown concentration with NaOH standard solution. You will determine the concentration of the phosphoric acid solution according to the titration curves.

REAGENTS AND APPARATUS

• Automatic titration system, pH compound electrodes, electromagnetic stirrer, 10 mL pipette, 100 mL beaker.
• $0.1 \text{ mol} \cdot L^{-1}$ sodium hydroxide (NaOH) standard solution, unknown phosphoric acid solution, distilled water.

PROCEDURES

Wash the automatic burette several times with distilled water, then rinse it with $0.100 \text{ mol} \cdot L^{-1}$ NaOH three times. Then fill the automatic burette with $0.100 \text{ mol} \cdot L^{-1}$

NaOH solution, and make sure the tip is filled with the NaOH solution and no bubbles are present in the tip. Calibrate the pH probe before starting the titrations. It needs a two point calibration using reference buffers of pH 4. 0 and pH 7. 0.

Pipet 10. 00 mL of the H_3PO_4 solution into a clean, dry 100 mL beaker. Add 40 mL of distilled water to the beaker. Place the 100 mL beaker on a magnetic stirrer and add a stirring bar. Using an apparatus clamp, position the pH electrode in the 100 mL beaker so that the probe is submerged in the solution but is not in any danger of being struck by the stir bar. Begin stirring the solution at a medium speed.

Begin the titration, observe the real time voltage change as NaOH is added. At the end of the titration, record the titration data. Repeat the titration three times.

After the experiment, wash the electrodes and put them away, turn off the power supply.

DATA TREATMENT

1. The potential vs. volume plots and their first derivative plots.
2. Titration results.

	1	2	3
First equivalence point V_{NaOH}			
Second equivalence point V_{NaOH}			
C_{H3PO4}			
average of C_{H3PO4}			
RSD			

QUESTIONS

1. Write out the balanced equation for the reaction of 1 mole of NaOH with 1 mole of H_3PO_4.

2. Write out the balanced equation for the reaction of 2 moles of NaOH with 1 mole of H_3PO_4.

3. How does buffering in a titration of a weak acid or weak base affect the shape of the titration curve when compared to the titration curve of strong acid with strong

base? Explain.

4. How to estimate K_{a1} and K_{a2} of phosphoric acid from titration plot?

2.32   The Redox of Potassium Ferricyanide on Glassy Carbon Electrode

AIMS

1. To learn the surface treatment of solid electrodes.

2. To master the use technology of cyclic voltammeter.

3. To understand the influence of scanning rates and concentration oncyclic voltammetry.

INTRODUCTION

Cyclic voltammetry (CV) is a versatile electroanalytical technique for the study of electroactive species. CV monitors redox behavior of chemical species within a wide potential range. The current at the working electrodeis monitored as a triangular excitation potential is applied to the electrode. The resulting voltammogram can be analyzed for fundamental information regarding the redox reaction. Cyclic voltammograms are the electrochemical equivalent to the spectra in optical spectroscopy.

In cyclic voltammetry, the electrode potential ramps linearly versus time in cyclical phases (Figure 2 – 46). The rate of voltage change over time during each of these phases is known as the experiment's scan rate ($V \cdot s^{-1}$). The potential is measured between the working electrode and the reference electrode, while the current is measured between the working electrode and the counter electrode. These data are plotted as current (i) versus applied potential (E, often referred to as just 'potential'). In Fig. 1, during the initial forward scan, an increasing oxidation potential is applied; thus the anodic current will, at least initially, increase over this time period assuming that there are oxidizable analytes in the system. At some point after the oxidation potential of the analyte is reached, the anodic current will decrease as the concentration of oxidizable analyte is depleted. If the redox couple is reversible, then during the reverse scan the oxidized analyte will start to be reduced, giving rise to a current of reverse polarity (cathodic current) to before. The more reversible the redox couple is, the more similar the reduction peak will be in shape to the oxidation peak. Hence, CV data can provide information about redox potentials and electrochemical reaction rates.

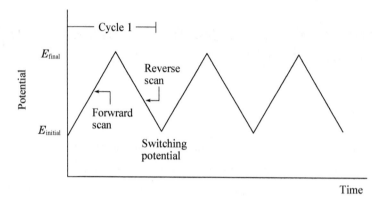

Figure 2 - 46   Cyclic voltammetry waveform.

Potassium ferricyanide is a bright red salt with a chemical formula $K_3Fe(CN)_6$. The $[Fe(CN)_6]^{3-}/Fe(CN)_6^{4-}$ redox couple is used as an example of an electrochemically reversible redox system used to study some basic concepts of cyclic voltammetry. $[Fe(CN)_6]^{3-}$ consists of a Fe^{3+} center bound in octahedral geometry to six cyanide ligands (shown in Figure 2 - 47). The iron is low spin and easily reduced to the related ferrocyanideion $[Fe(CN)_6]^{4-}$, which is a ferrous (Fe^{2+}) derivative.

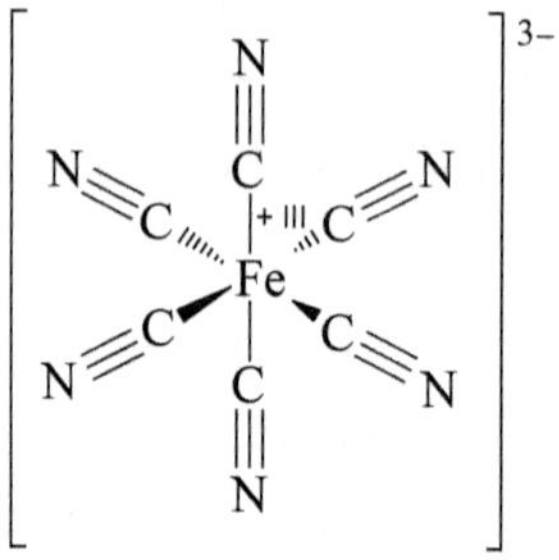

Figure 2 - 47   The structure of $[Fe(CN)_6]^{3-}$

As the potential is scanned positively and is sufficiently positive to oxidize $[Fe(CN)_6]^{4-}$, the anodic current is due to the electrode process

$$[Fe(CN)_6]^{4-} \longrightarrow [Fe(CN)_6]^{3-} + e \qquad (2\text{-}32\text{-}1)$$

The electrode acts as an oxidant and the oxidation current increases to a peak. The concentration of $[Fe(CN)_6]^{4-}$ at the electrode surface depletes and the current then decays. As the scan direction is switched to negative, for the reverse scan the potential is still sufficiently positive to oxidize $[Fe(CN)_6]^{4-}$, so anodic current continues even though the potential is now scanning in the negative direction. When the electrode becomes a sufficiently strong reductant, $Fe(CN)_6^{-3}$, which has been formed adjacent to the electrode surface, will be reduced by the electrode process:

$$[Fe(CN)_6]^{3-}+e\longrightarrow[Fe(CN)_6]^{4-} \tag{2-32-2}$$

resulting in a cathodic currentwhich peaks and then decays as $Fe(CN)_6^{-3}$ in the solution adjacent to the electrode is consumed.

In the forward scan $[Fe(CN)_6]^{3-}$ is electrochemically generated from $[Fe(CN)_6]^{4-}$ (anodic process) and in the reverse scan this $[Fe(CN)_6]^{3-}$ is reduced back to $[Fe(CN)_6]^{4-}$ (cathodic process). Note that the technique of CV rapidly generates various oxidation states.

The quantities of note a CV plot are the anodic peak current i_{pa}, cathodic peak current i_{pc}, anodic peak potential E_{pa}, and cathodic peak potential E_{pc}. Measuring i_p does involve the extrapolation of the base-line current.

A redox couple in which half reactions rapidly exchange electrons at the working electrode are said to be electrochemically reversible couples. The formal reduction potential $E^{o'}$ (different from E^o) for such a reversible couple is the mean of E_{pa} and E_{pc} and the i_{pa} and i_{pc} are very close in magnitude:

$$E^{o'}=\frac{E_{pa}+E_{pc}}{2} \tag{2-32-3}$$

The number of electrons involved in the redox reaction for a reversible couple is related to the difference of peak potentials by:

$$E_{pa}-E_{pc}=\frac{59}{n}mV \tag{2-32-4}$$

For slow electron transfers at the electrode surface, i. e. irreversible processes, the difference of peak potentials widen. The peak current in reversible systems for the forward scan is given by Randles-Sevcik equation:

$$i_{pc}=2.69\times10^8 n^{3/2}AD^{1/2}v^{1/2}C \tag{2-32-5}$$

where, i_{pc} is peak current in the unit of A, n is electrons involved in the reaction process, A is the electrode area in the unit of m^2, D is the diffusion coefficient in the unit of $m^2 \cdot s^{-1}$, C is concentration in the unit of $mol \cdot L^{-1}$, and v is scan rate in the unit of $V \cdot s^{-1}$. Thus i_{pc} increases with square root of v and is directly proportional to concentration of the species. The values of i_{pa} and i_{pc} are very similar for a one step reversible couple leading to their ratio to be unity. Ratio of peak currents may differ from unity if the reactions coupled to other electrode processes.

REAGENTS AND APPARATUS

• Electrochemical station, glassy carbon working electrode (diameter 3 mm), platinum auxiliary electrode, Saturated calomel reference electrode, nitrogen cylinder, oxygen absorber, polishing material.

• 10 mmol • L^{-1} $K_4Fe(CN)_6$ in 1.0 mol • L^{-1} potassium nitrate (KNO_3) solution, 1.0 mol • L^{-1} KNO_3 (matrix), unknown $K_4Fe(CN)_6$ in 1.0 mol • L^{-1} KNO_3 solution, 1 000 ppm $K_4Fe(CN)_6$ in 1.0 mol • L^{-1} KNO_3 solution, distilled water.

PROCEDURES

The carbon working electrode was polished using alumina slurry (0.05 micron alumina powder) on soft lapping pads and rinsed well with distilled water before experimentation to have a fresh working surface. Fill the cell with the 1.0 mol • L^{-1} KNO_3 supporting electrolyte. The supporting electrolyte is to control electrode potentials, eliminate the transport of electroactive species from migrating in the electric field gradients, maintain constant ionic strength and maintain constant pH. This ensured a diffusion controlled electrode process. Direct reduction potential of iron was then determined.

Deoxygenate the solution by purging with oxygen free nitrogen for approximately 15 min. Turn off the purging but maintain an envelope of nitrogen over the solution. Set the scan parameters suggested below, they may be changed appropriately.

CV Parameters:

Initial E (mV)	−100	Scan rate (mV • s^{-1})	100
High E (mV)	+600	Sensitivity (A • V^{-1})	10^{-4}
Low E (mV)	−100	Quiet time (s)	2
Final E (mV)	−100	Number of scans	2

Excite the working electrode with the potential scan and obtain the background CV of the supporting electrolyte solution. (Save all data in named files.)

Change the solution, fill the cell with 10 m mol • L^{-1} $K_4Fe(CN)_6$ in 1.0 mol • L^{-1} KNO_3 solution. Following the same procedure as above, obtain the CV of the redox couple.

Using the same solution and obtain CV's at the following scan rates: 20, 50, 75, 125, 150, 175, and 200 mV • s^{-1}. Between each scan, pass nitrogen through the cell to restore initial conditions and allow the system to acquire quiescence before applying the excitation potential scan.

Determine the concentration of the unknown by the standard addition method.

DATA TREATMENT

1. Present the cyclic voltammograms of a set of 10 mmol • $L^{-1}K_4Fe(CN)_6$ with

different scan rates overlapped on the same figure, the plot after addition with unknown $K_4Fe(CN)_6$ solution, and the background.

2. Plot i_{pc} versus $v^{1/2}$ and i_{pa} versus $v^{1/2}$.

3. Plot E_p values versus scan rate (v) for the above runs-comment on the reversibility.

4. Verify the formal standard electrode potential of Fe(II)/Fe(III) system and the number of electrons involved in the half reaction, n, from appropriate plots.

5. Determine the diffusion coefficients of the two species.

6. Using the standard addition method to determine the concentration of the unknown.

QUESTIONS

1. For a simple electrode reaction process, what are the important basises for judging whether the electrode reaction is a reversible system?

2. What issues should be paid attention to in the pretreatment of electrodes?

3. Explain the possible reaction mechanism of potassium ferricyanide on the electrode by cyclic voltammetry.

2.33 Analysis of Alcohol in Alcoholic Beverages by Gas Chromatography

AIMS

1. To understand the technique of gas chromatography and be able to operate a modern GC instrument.

2. To employ GC analysis to determine the composition of a commercially available alcoholic beverage.

INTRODUCTION

Chromatography is a laboratory technique for the separation of a mixture. The mixture is dissolved in a fluid (gas or liquid) called the mobile phase, which carries it through a system (a column, a capillary tube, a plate, or a sheet) on which is fixed a material called the stationary phase. The different constituents of the mixture have different distributions between two phases. The distribution ratio is determined by the physical and chemical characteristics of the molecules and by the operating conditions such as the temperature. As the mobile phase flows over the stationary phase, the solute molecules are selectively retarded and separated out from the matrix of the sample.

In gas chromatography the separation is carried out by injecting a liquid sample into a stream of gas which flows at a selected temperature through a column which is packed with a uniform solid support material (often coated with a liquid stationary phase). The injection temperature is usually at about 50 °C above the boiling point of the least volatile component, and so the sample is readily vaporized upon injection. The column may be maintained at a constant temperature throughout the separation, or the temperature may be progressively increased (programmed) to improve the characteristics of the separation. The presence of each component emerging from the column may be recorded and quantified by measuring some physical property (thermal conductivity, density, flame ionisation, etc) of the carrier gas at the end of the column (see Figure 2 – 48). One of the most common detection systems is based on the flame ionization detector (FID) that responds to most organic compounds and is very sensitive.

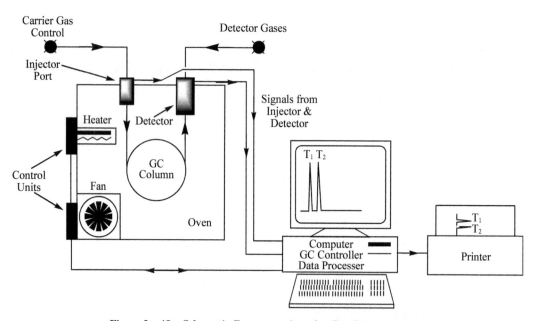

Figure 2 – 48   Schematic Representation of a Gas Chromatograph

Under ideal conditions the chromatogram shows a series of separate peaks, each corresponding to a single component in the mixture. These quence in which the peaks are eluted from the column corresponds to the degree of retardation (retention) of each component on the column. For analytical purposes, each peak has two important characteristics: retention time (t_r) and peak area (A).

Retention time is the time taken for a component to elute from the column depending upon the chemical and physical interactions between the solute and the stationary phase, upon the vapour pressure of the solute, and upon the temperature of the column.

The peak area (A) of each peak is related to the amount (or concentration) of the substance which produces the peak. Thus the peak areas from the chromatograms can be utilised to determine the concentration of the respective analytes in a mixture. However, the use of peak areas is complicated by the fact that the area is also dependent upon certain operational parameters, including the response characteristics of the detector with respect to specific molecules.

In order to overcome this problem, an internal standard method is employed. A known amount of an internal standard (IS) is added to a standard sample and comparisons of peak areas are made between the known amounts of analyte (X) and IS. The response of the detector to the analyte X in the presence of the IS is termed the detector response factor (F). The DRF is obtained using:

$$F=\frac{A_x}{A_{IS}}\times\frac{c_{IS}}{c_X}$$ (2-33-1)

Where F is detector response factor, A is the peak area of either the internal standard or analyte, c is the concentration of either the internal standard or analyte.

The F factor can be utilized to determine the concentration of the analyte X in an unknown sample by first adding a known amount of IS into that sample prior to analysis and then suitably rearranging Equ. (2-33-1). In this experiment the analysis of a mixture of alcohols is accomplished using propan – 1 – ol as an internal standard. The application of this principle can be illustrated by an example:

A mixture of ethanol and propan – 1 – ol is made up accurately by mixing 400 μL of an 80 mg% ethanol solution with 1 000 μL of a 25 mg% propan – 1 – ol solution. In all subsequent preparations identical solutions are made i. e. 400 : 1 000 μL ethanol : propan – 1 – ol additions. Hence dilution factors are standardized and thus the c_{IS} : c_X ratio from Equ. (2-33-1) is simply 25 : 80. The chromatogram of this mixture produces two peaks, with associated peak areas and hence the detector response factor (F) for ethanol relative to propan – 1 – ol can be determined using Equ. (2-33-1). When the F is known, a mixture of ethanol and propan – 1 – ol may be analysed in the same manner by measuring the ratio of the peak areas and the amount of ethanol determined by rearranging Equ. (2-33-1) appropriately.

REAGENTS AND APPARATUS

• Gas Chromatograph, microsyringes, appropriate volumetric glassware, glass sample vials.

 • 0. 8 mg • mL^{-1} and 2 mg • mL^{-1} ethanol, 0. 25 mg • mL^{-1} Propan－1－ol, deionised water, commercial beverage.

PROCEDURES

Before the experiment, check the following settings on the GC firstly: Injector and detector temperature ＞200 ℃; Column temperature 100 ℃ for isothermal analysis.

1. Determination of the detector response factor (F) for ethanol relative to propan－1-ol

All samples and standards are aqueous solutions. Do not use neat alcohols in this experiment.

Prepare a calibration mixture in triplicate as follows: pipette 1 000 μL of the 0. 25 mg • mL^{-1} propan－1－ol internal standard solution into a clean dry clip top vial, followed by 400 μL of the 0. 8 mg • mL^{-1} ethanol standard solution. Cap the vials until required to prevent loss of ethanol.

In an identical manner, prepare, in triplicate, a second calibration mixture by taking up 1 000 μL of the 0. 25 mg • mL^{-1} propan－1－ol solution and 400 μL of the 2 mg • mL^{-1} ethanol solution. Cap the vials until required to prevent loss of ethanol.

Take 1 μL of each mixture in turn, and inject it onto the column through the septum on the top of the instrument. Apply a positive action to the syringe plunger, and at the time of the injection activate the integrator by pressing "run" or "start" as appropriate. Withdraw the syringe from the septum after about 3 seconds. Each alcohol will produce a peak that is characterized by a retention time (t_r) and a peak area (A). Record the retention times and peak areas for ethanol and propan－1－ol in accurately prepared calibration mixtures. Calculate the F for ethanol relative to propan－1－ol for each injection, and tabulate your results. Flush out the syringe 3 or 4 times with a supply of the next sample, before injecting a sample onto the column.

Calculate the mean value of the F after rejecting any outliers. For a 95% degree of confidence, results should be within＋/－2% of the mean. At least 4 of the DRF values should be in agreement.

Calculate the Range Error 'E_r' of the accepted values using Equ. 2-33-2:

$$E_r = \frac{(Xhighest\text{-}Xlowest)}{\overline{X}} \times 100 \qquad (2\text{-}33\text{-}2)$$

2. Analysis of ethanol in a sample of a beer or wine

Take approximately 20 mL of the alcoholic beverage and place it in a capped container. Ensure that it is fullydegassed using an ultrasonic bath if necessary. Discard the excess after use. From the degassed sample prepare a diluted solution in triplicate by taking 2 mL of the degassed drink and diluting it to 50 mL using distilled water to give a dilution factor of 1 : 25.

The analytical samples for analysis are now prepared as before by taking up 1 000 μL of the 0. 25 mg • mL^{-1} propan – 1 – ol of internal standard followed by 400 μL of the diluted drink. Accurate addition by electronic pipette is required, and the samples should be made up in triplicate and stored in stoppered vials as before to ensure minimal loss of ethanol vapours. Inject separate 1 μL samples onto the column and record the retention times and peak areas for ethanol and propan – 1 – ol, respectively. Using the detector response factors measured earlier, calculate the concentration of ethanol in the unknown, and report your results in a tabulated format. Calculate the range error for the analyses.

DATA TREATMENT

1. The GC spectra of samples.
2. Determination of F for ethanol relative to propan – 1 – ol.

	1	2	3	4
t_r of ethanol				
A of ethanol				
t_r of propan – 1 – ol				
A of propan – 1 – ol				
F				
E_r				

3. Analysis of ethanol in the sample.

	t_r			A	
ethanol					
propan – 1 – ol					
Concentration of ethanol					
E_r					

QUESTIONS

1. Briefly explain the mechanisms of the separative process that is occurring within the column in a GC system. What factors can be altered to affect the separation?

2. Describe a method by which non-volatile alcohols or carboxylic acids may beanalysed by GC.

2.34 The Separation and Quantification of Food Additives in Drinks Using High Performance Liquid Chromatography

AIMS

1. To learn how to operate a modern HPLC instrument.

2. To use HPLC analysis to determine the type and quantification of some commercial food additives.

INTRODUCTION

The term chromatography was first employed in 1903 by the Russian botanist and chemist Michael Tswett to describe the separation of coloured plant pigments using a packed columnar bed of chalk (the stationary phase) and aqueous solvents (the mobile phase). Chromatographic separations occur as a result of the differing specific affinities of each analyte for the mobile and stationary phases, respectively. The specific affinities are in turn dependent upon the chemical and physical characteristics of each of the individual components with respect to the mobile and stationary phases that are being employed.

Liquid chromatography is now used extensively in research, development and production environments. However, initial chromatographic separative techniques employed relatively coarse (diameters ~ > 150 mm) and irregular shaped adsorbent materials packed into large columns (up to 5 cm in diameter and 2 m in length). These columns utilized atmospheric pressure to force the eluant through the stationary phase (so-called gravity-fedcolumns). The mobile phase flow rates in these columns were very slow, requiring large quantities of solvent to effect separations and as a consequence, these systems were laborious to use.

Van Deemter determined that the separation efficiency of chromatographic columns was directly related to the size of the particulate packing material. However, decreasing the dimensions of the packing material results in drastically reduced mobile phase flow rates when using gravity fed techniques. The solution to this problem was to apply pressure to force the mobile phase through the stationary phase (commonly called flash chromatography). The development of reliable high pressure pumping systems has ensured that this technique is now prevalent in laboratories.

Modern analytical liquid chromatographic systems feature metal columns, typically 10—25 cm long with an internal diameter of 4. 6 mm, and packed with a regular shaped stationary phase with dimensions in the range of 3—10 mm. Several classes of stationary phase are used, dependent upon the type of analyte under investigation. The majority of stationary phases are based upon silica. Silica can be used directly as a polar stationary phase, in which case the mobile phases employed are non-polar in nature (normal phase liquid chromatography). Alternatively, modified bonded phase silica packings can employed that are produced by reacting an organochlorosilane with the silanol (R_3Si—OH) functionalities present on the silica gel surface to form silyl ether linkages (Figure 2 – 49). The most common class of bonded phases are those featuring linear alkyl units, such as the octadecylsilyl (ODS) $C_{18}H_{37}$ unit. These bonded phases are non-polar in nature and polar solvents are routinely used as the mobile phase (referred commonly to as reverse phase liquid chromatography).

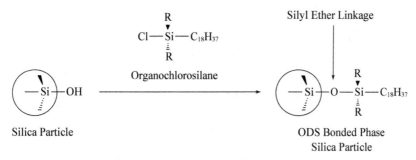

Figure 2 – 49 Silyl ether synthesis to form ODS bonded phase column packings.

A schematic diagram of a modern HPLC system is shown in Figure 2 – 50. It is now common for the provision of two or more solvent reservoirs in order to allow the mobile phase composition to be altered during the course of the analysis. Prior to analysis, the mobile phase solvents must be filtered to remove particulate material and also thoroughly degassed to remove dissolved gases. Solvent degassing is accomplished by bubbling vigorously an inert gas (helium) through the solvent or by ultrasonication. The mobile phase composition (and therefore the polarity) is regulated by a solvent control unit. This unit can be programmed to deliver the high pressure pump with either (i) a constant mobile phase composition over the whole duration of the analysis (isocratic elution) or alternatively (ii) a solvent composition that changes during the separation process (gradient elution). Gradient elution conditions are employed when it is necessary to change the mobile phase polarity during the analysis in order to improve the efficiency of the separation.

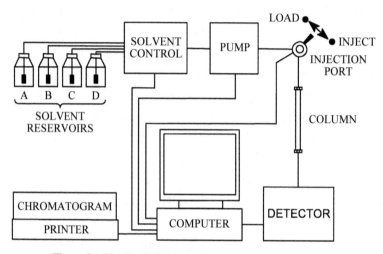

Figure 2 - 50   Typical Layout for a modern HPLC System

The mobile phase is pumped under constant pressure (up to ~48 MPa) onto the column *via* a sample injection port. Sample injection port systems are designed to introduce the sample onto the head of the column without introducing air into the system and to minimize disruption to the mobile phase flow. Manual sample injection ports utilize a sample loop that will hold a defined volume (typically 550 mL). These injection systems operate by first loading the sample into the sample loop using a suitable syringe while the mobile phase is guided directly onto the column (the load position). The injection port is then swiftly switched to the inject position. Modern HPLC systems often feature a programmable auto injection system—these allow for repeat injections to be made from one sample or alternatively for large numbers of samples to be injected in sequence. Such automated injection systems allow for overnight analyses to be performed and have proven invaluable in environmental and pharmaceutical analysis where the number of samples produced are large. In both of the automated and manual injection systems, the mobile phase flow is directed through the sample loop and flushes the sample onto the head of the column.

Upon eluting from the column, the separated analytes are detected using a suitable detection system. Spectrophotometric measurements are the most popular detection systems and the Beer-Lambert law can be employed to relate absorbance to analyte concentration. The most commonly employed detection system is the single wavelength ultraviolet (UV) detector—the majority of organic analytes absorb in the UV region of the electromagnetic spectrum. However, numerous detector modes have been employed successfully in conjunction with HPLC based separations, ranging from refractive index (RI) detectors to mass spectrometric (MS) systems. The detector response is subsequently converted into an electrical output that is used to generate a plot of detector response *vs* time—otherwise known as a chromatogram. An ideal chromatogram will reveal a series of baseline separated peaks, each peak corresponding to a single component in the original mixture. The data that is obtained directly from chromatograms (analyte retention times, peak areas, peak widths) can be utilized to assess the separative performance of the separation system being used and also to determine the concentration of analytes in sample mixtures.

The capacity (or retention) factor k is used to express the retentive abilities of a column and describes the equilibrium distribution occurring on the column, taking into account experimental conditions. Optimum k values lie between 2 to 6. The capacity factor is calculated using:

$$k = \frac{t_r - t_0}{t_0} \tag{2-34-1}$$

Where k is capacity factor, t_r is retention time of analyte, and t_0 is retention time of unretained component (typically a solvent).

The separation efficiency of a column can be assessed by considering the number of equilibrium/separation steps that is occurring for a particular analyte. It is common for current HPLC columns to exhibit efficiencies in excess of 50, 000 plates per metre. The separation efficiency is described using the term: effective number of theoretical plates or n_{eff} and can be determined using the following equation:

$$n_{neff} = 5.54 \left(\frac{t_r'}{w_h}\right)^2 \tag{2-34-2}$$

Where n_{eff} is effective number of the oretical plates, t_r' isadjusted retention time of analyte $(t_r' = t_r - t_0)$, w_h is peak width at half height.

Chromatographic separations are assessed by the resolution between the peaks of separated components, ie, adjacent peaks should be sufficiently resolved in order to allow accurate peak area determinations to be made. In order to meet this criteria, there should be baseline separation and no co-elution of the analytes. The effectiveness of a separation between two analytes is assessed by the resolution R. This factor can be obtained using:

$$R = 2 \frac{t_{rB} - t_{rA}}{w_B + w_A} \tag{2-34-3}$$

Where R is resolution, t_r is retention times of analytes A and B, w is peak width of analytes A and B.

For an ideal separation, R values should lie between 1. 2—1. 5, whereas >1.8 indicates analysis times that are too long with potential band broadening processes occurring.

The peak areas from the chromatograms can be utilized to determine the concentration of the respective analytes in a mixture. However, the peak areas cannot be used directly—not all the analytes will exhibit the same response characteristics as the detection method being employed. In order to overcome this problem, an internal standard method is employed. A known amount of an internal standard (IS) is added to a standard sample and comparisons of peak areas are made between the known amounts of analyte (X) and IS. The response of the detector to the analyte X in the presence of the IS is termed the detector response factor (F), which is obtained using:

$$F = \frac{A_x}{A_{IS}} \times \frac{c_{IS}}{c_X} \qquad (2\text{-}34\text{-}4)$$

Where F is detector response factor, A is peak area of either the internal standard or analyte, c is concentration of either the internal standard or analyte.

The F factor can be utilized to determine the concentration of the analyte X in an unknown sample by first added a known amount of IS into that sample prior to analysis and then suitably rearranging the equation (2-34-4).

As consumers pay more and more attention to food safety, the use of various additives in food, especially synthetic additives, has attracted more and more attention. Benzoic acid, sorbic acid, saccharin sodium and other additives are widely used in drinks, pickles, preserves and other foods. The long-term excessive consumption of these additives has certain harm to the human body. The scope and the maximum limit of use of these additives are clearly stipulated in China's "Hygienic standard for the use of food additives" (GB 2762—2014): Benzoic acid and its sodium salt$\leqslant$1. 0 g • kg^{-1}, sorbic acid and its potassium salt$\leqslant$0. 5 g • kg^{-1}, and saccharin sodium is prohibited.

In this experiment, you need to demonstrate the basic principles of the HPLC technique by optimising the mobile phase composition for the separation of food additives (see Figure 2 – 51a – c) and to determine the composition and content of these additives in a commercial drink.

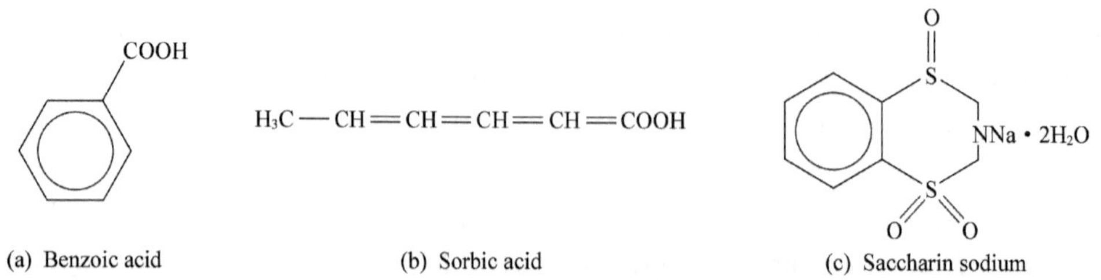

(a) Benzoic acid   (b) Sorbic acid   (c) Saccharin sodium

Figure 2 – 51 Some food additivestructures

REAGENTS AND APPARATUS

• Programmable HPLC System: solvent degasser, HPLC pump, auto injector, 5 mm ODS column: 15 cm ×4. 6 mm i. d. , column oven, UV detector, appropriate volumetric glassware.

• HPLC grademethanol, HPLC grade water, 0. 02 mol • L^{-1} ammonium acetate solution, 20 g • L^{-1} sodium bicarbonate solution, standard mixture containing 1 : 1 : 1 of benzoic acid : sorbic acid : saccharin sodium, commercial fruit drink.

PROCEDURES

1. Analysis of standard solutions and method development

Prepare an internal standard stock solution by weighing out accurately 100 mg of diethyl phthalate into a 250 mL graduated flask. Carefully add HPLC grade water until the stock solution volume is just below the 250 mL volumetric mark. Shake the mixture and sonicate the solution thoroughly to ensure complete dissolution. Allow the solution to stand for approximately 1 minute. Carefully add water in a dropwise fashion using a clean Pasteur pipette up to the mark (bottom of the solutions' meniscus in line with the volumetric mark).

Prepare a standard solution by weighing out accurately 60 mg of the standard mixture that contains equal amounts of benzoic acid, sorbic acid, and saccharin sodium into a 50 mL graduated flask. Using 20 g • L^{-1} sodium bicarbonate solution as the solvent, prepare this solution for analysis following the same procedure described for the preparation of the internal standard solution.

Prior to HPLC analysis, the UV spectral characteristics of the components in the standard mixture must be considered in order to select the optimum detection wavelength. Examine the UV spectra of the individual components that are provided on the information board adjacent to the HPLC instrument. Carefully select the wavelength that will give the optimum detector response for all the components— verify your selection with the demonstrator before altering the wavelength setting on the UV detector.

In order to determine the optimum mobile phase composition for the analysis of the analgesic mixture, the HPLC system should be programmed to pump a series of different solvent mixtures through the column. Program the HPLC system to pumpinitially 5% methanol: 90% of 0. 02 mol • L^{-1} ammonium acetate(aq) through the column. A period of 10 minutes is required between adjusting the mobile phase compositions to ensure that the column is in an equilibrated state prior to analysis.

Decrease the ammonium acetate component of the mobile phase in 5% reduction down to a value of 80% and analyze the standard mixture under each of these conditions. Record the retention times and peak areas that are obtained for each component using the different mobile phase conditions. The elution order of the components is: i) benzoic acid, ii) Sorbic acid, iii) Saccharin sodium. Identify the optimum mobile phase composition for the analysis of the standard solution that gives the best compromise between resolution and separation time. Use the chromatographic data obtained for each of the mobile phase conditions to calculate the capacity factor, efficiency and resolution in order to select the optimum mobile phase composition. Note to ask the demonstrator to help you determine the t_0 value in order that you can calculate the capacity factor values.

Using the chromatographic data obtained from the analysis that employed the optimum mobile phase composition, determine the detector response factors (F) (see Equ. 4) for benzoic acid, sorbic acid, and saccharin sodium with respect to the internal standard, diethyl phthalate.

2. Sample analysis of commercial fruit drink

Prepare a solution of the commercial fruit drink as follows. Accurately weigh 5.00 g fruit drink, and control the pH value to 7 by 1 : 1 aqueous ammonia. Add the solution into a 50 mL graduated flask, and use HPLC grade water to the volumetric mark. After centrifugation, the supernatant is filtered by 0.45 mm filter membrane. Analyze the fruit drink solution in duplicate using the optimum mobile phase conditions.

Employ the peak areas from the commercial fruit drink chromatogram and the F values obtained from the standard solution analysis at the optimum mobile phase conditions to calculate the fruit drink composition.

DATA TREATMENT

1. The LC spectra of samples.
2. Analysis of standard solutions and method development.

	diethyl phthalate		benzoic acid		sorbic acid		saccharin sodium	
	t_r	A	t_r	A	t_r	A	t_r	A
mobile 5 : 95								
mobile 10 : 90								

(Continued)

	diethyl phthalate		benzoic acid		sorbic acid		saccharin sodium	
	t_r	A	t_r	A	t_r	A	t_r	A
mobile 15:85								
mobile 20:80								
k								
n_{eff}								
R								
F								

3. Sample analysis of commercial fruit drink.

	diethyl phthalate		benzoic acid		sorbic acid		saccharin sodium	
	t_r	A	t_r	A	t_r	A	t_r	A
1								
2								
C_1								
C_2								
C_{mean}								

QUESTIONS

1. Briefly explain the mechanism of the separative process that is occurring within the column of a HPLC system.

2. Describe the mode of operation of a column that is packed with size exclusion media for the separation of polystyrene oligomers.

3. What is the role of guard columns in modern HPLC systems?

2. 35   Thermal behavior and decomposition of copper sulfate pentahydrate

AIMS

1. To learn the principle of thermogravimetric analysis and the basic structure of the thermal analyzer.

2. To know how to operate the thermogravimetric analyzer.

3. To determine the differential thermal spectra of copper sulfate crystals, and analyze the chemical changes during the heating process.

INTRODUCTION

Thermogravimetric analysis (TGA) is a method of thermal analysis in which the mass of a sample is measured over time as the temperature changes. This measurement provides information about physical phenomena, such as phase transitions, absorption, adsorption and desorption; as well as chemical phenomena including chemisorptions, thermal decomposition, and solid-gas reactions (e. g. , oxidation or reduction).

TGA is conducted on an instrument referred to as a thermogravimetric analyzer. A thermogravimetric analyzer continuously measures mass while the temperature of a sample is changed over time. Mass, temperature, and time are considered base measurements in thermogravimetric analysis while many additional measures may be derived from these three base measurements.

A typical thermogravimetric analyzer consists of a balance, a furnace, a temperature programmer, and a recording system (Figure 2 – 52). One important part of a thermogravimetric analyzer is a precision balance, which looks like a sample pan located inside a furnace with a programmable control temperature. The balance is basically a high-quality analytical balance, and it has high accuracy, fine reproducibility, great seismic performance, and good reactivity. Another feature is the temperature programmer, which consists of two thermocouples connected to a voltmeter. One thermocouple is placed in an inert material such as Al_2O_3, while the other is placed in a sample of the material under study. As the temperature increases, there will be a brief deflection of the voltmeter if the sample is undergoing a phase transition. This occurs because the input of heat will raise the temperature of the inert substance, but be incorporated as latent heat in the material changing phase.

The temperature is generally increased at constant rate (or for some applications the temperature is controlled for a constant mass loss) to incur a thermal reaction. The thermal reaction may occur under a variety of atmospheres including: ambient air, vacuum, inert gas, oxidizing/reducing gases, corrosive gases, carburizing gases, vapors of liquids or "self-generated atmosphere"; as well as a variety of pressures including: a high vacuum, high pressure, constant pressure, or a controlled pressure.

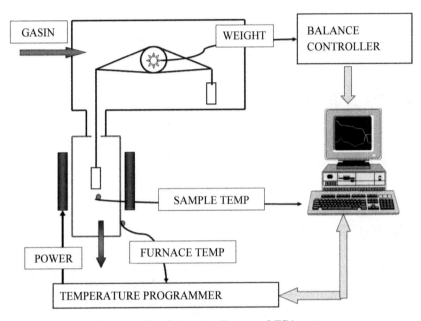

Figure 2 - 52　Schematic diagram of TGA system

The thermogravimetric data collected from a thermal reaction is compiled into a plot of mass or percentage of initial mass on the y axis versus either temperature or time on the x-axis. This plot, which is often smoothed, is referred to as a TGA curve. The first derivative of the TGA curve (the DTG curve) may be plotted to determine inflection points useful for in-depth interpretations as well as differential thermal analysis.

There are several types of thermogravimetric analysis: One is isothermal or Static Thermogravimetry. In this technique, the sample weight is recorded as function of time at constant temperature. Another is quasistatic thermogravimetry: In this technique the sample is heated to a constant weight at each of increasing temperatures. The third is dynamic thermogravimetry, in which the sample is heated in an environment, whose temperature is changed in linear manner.

A TGA can be used for materials characterization through analysis of characteristic decomposition patterns. It is an especially useful technique for the study of polymeric materials, including thermoplastics, thermosets, elastomers, composites, plastic films, fibers, coatings, paints, and fuels.

The mass of crystal water in hydrates can be measured using thermogravimetry. When copper(II) sulfate pentahydrate ($CuSO_4 \cdot 5H_2O$) is heated, it decomposes to the dehydrated form. The waters of hydration are released from the solid crystal and form water vapor. The hydrated form is medium blue crystal, and the dehydrated solid is white powder. $CuSO_4 \cdot 5H_2O$ loses its water of crystallization in several steps.

The balanced equation is:

$$CuSO_4 \cdot 5H_2O \text{ (s)} \longrightarrow CuSO_4 \text{(s)} + 5H_2O\text{(g)} \qquad (2\text{-}35\text{-}1)$$

REAGENTS AND APPARATUS

• Comprehensive thermal analyzer, ananlytical balance, Al_2O_3 crucibles, crucible tweezers, ladles.

• Copper sulfate pentahydrate ($CuSO_4 \cdot 5H_2O$).

PROCEDURES

Adjust the null position of the balance. Put the crucible on the balance and record its value P_1 to four digits after decimal point. Then add the sample into 1/2 or 2/3 of the crucible, and then weight the crucible again and record the initial weight value P_2.

Place the crucible in the hanging plate of the heating furnace. Open cooling system (usually water cooling) and enter the inert gas (usually high pure nitrogen). Set the temperature from room temperature to 350 ℃, select the heating rate at 20 K • min^{-1} (if it is automatically recorded, set the speed of the paper and open the recorder).

Start the electric furnace power, so that heating is at a given speed. Observe the thermometer, open the balance, read and record the weight value at a certain time (if it is a automatic recorder, then observe the TG curve regularly and record the mass and temperature values), record the final weight value P_3.

After the experiment, cut off the power supply firstly. When the furnace temperature is less than 100 ℃, the cooling system can be cut off.

According to the mass recorded, calculate the weight loss according to the formula:

$$WL = \frac{P_3 - P_2}{P_2 - P_1} \times 100\% \qquad (2\text{-}35\text{-}2)$$

DATA TREATMENT

1. The TGA plot of $CuSO_4 \cdot 5H_2O$.
2. Analyze the reason of the change on the TGA plot of $CuSO_4 \cdot 5H_2O$.
3. Data record.

	Mass (mg)
crucible P_1	
crucible+sample (beginning) P_2	
crucible+sample (final) P_3	
weight loss WL	

QUESTIONS

1. How to choose the experimental conditions to make the multi-step decomposition TG curve clear and distinguishable?

2. What are the factors of the measurementaccuracy and what should we do to improve the accuracy in the experiment?

3. What are the characteristics of comprehensive thermal analysis? Try to sum up some rules.

3 Appendix

3.1 Chemical Elements Listed by Atomic Mass

Atomic Mass	Name chemical element	Symbol	Atomic number
1. 007 9	Hydrogen	H	1
4. 002 6	Helium	He	2
6. 941	Lithium	Li	3
9. 012 2	Beryllium	Be	4
10. 811	Boron	B	5
12. 010 7	Carbon	C	6
14. 006 7	Nitrogen	N	7
15. 999 4	Oxygen	O	8
18. 998 4	Fluorine	F	9
20. 179 7	Neon	Ne	10
22. 989 7	Sodium	Na	11
24. 305	Magnesium	Mg	12
26. 981 5	Aluminum	Al	13
28. 085 5	Silicon	Si	14
30. 973 8	Phosphorus	P	15
32. 065	Sulfur	S	16
35. 453	Chlorine	Cl	17
39. 098 3	Potassium	K	19
39. 948	Argon	Ar	18
40. 078	Calcium	Ca	20
44. 955 9	Scandium	Sc	21
47. 867	Titanium	Ti	22
50. 941 5	Vanadium	V	23
51. 996 1	Chromium	Cr	24

Atomic Mass	Name chemical element	Symbol	Atomic number
54. 938	Manganese	Mn	25
55. 845	Iron	Fe	26
58. 693 4	Nickel	Ni	28
58. 933 2	Cobalt	Co	27
63. 546	Copper	Cu	29
65. 39	Zinc	Zn	30
69. 723	Gallium	Ga	31
72. 64	Germanium	Ge	32
74. 921 6	Arsenic	As	33
78. 96	Selenium	Se	34
79. 904	Bromine	Br	35
83. 8	Krypton	Kr	36
85. 467 8	Rubidium	Rb	37
87. 62	Strontium	Sr	38
88. 905 9	Yttrium	Y	39
91. 224	Zirconium	Zr	40
92. 906 4	Niobium	Nb	41
95. 94	Molybdenum	Mo	42
98	Technetium	Tc	43
101. 07	Ruthenium	Ru	44
102. 905 5	Rhodium	Rh	45
106. 42	Palladium	Pd	46
107. 868 2	Silver	Ag	47
112. 411	Cadmium	Cd	48
114. 818	Indium	In	49
118. 71	Tin	Sn	50
121. 76	Antimony	Sb	51
126. 904 5	Iodine	I	53
127. 6	Tellurium	Te	52
131. 293	Xenon	Xe	54
132. 905 5	Cesium	Cs	55

(**Continued**)

Atomic Mass	Name chemical element	Symbol	Atomic number
137. 327	Barium	Ba	56
138. 905 5	Lanthanum	La	57
140. 116	Cerium	Ce	58
140. 907 7	Praseodymium	Pr	59
144. 24	Neodymium	Nd	60
145	Promethium	Pm	61
150. 36	Samarium	Sm	62
151. 964	Europium	Eu	63
157. 25	Gadolinium	Gd	64
158. 925 3	Terbium	Tb	65
162. 5	Dysprosium	Dy	66
164. 930 3	Holmium	Ho	67
167. 259	Erbium	Er	68
168. 934 2	Thulium	Tm	69
173. 04	Ytterbium	Yb	70
174. 967	Lutetium	Lu	71
178. 49	Hafnium	Hf	72
180. 947 9	Tantalum	Ta	73
183. 84	Tungsten	W	74
186. 207	Rhenium	Re	75
190. 23	Osmium	Os	76
192. 217	Iridium	Ir	77
195. 078	Platinum	Pt	78
196. 966 5	Gold	Au	79
200. 59	Mercury	Hg	80
204. 383 3	Thallium	Tl	81
207. 2	Lead	Pb	82
208. 980 4	Bismuth	Bi	83
209	Polonium	Po	84
210	Astatine	At	85
222	Radon	Rn	86

(**Continued**)

Atomic Mass	Name chemical element	Symbol	Atomic number
223	Francium	Fr	87
226	Radium	Ra	88
227	Actinium	Ac	89
231. 035 9	Protactinium	Pa	91
232. 038 1	Thorium	Th	90
237	Neptunium	Np	93
238. 028 9	Uranium	U	92
243	Americium	Am	95
244	Plutonium	Pu	94
247	Curium	Cm	96
247	Berkelium	Bk	97
251	Californium	Cf	98
252	Einsteinium	Es	99
257	Fermium	Fm	100
258	Mendelevium	Md	101
259	Nobelium	No	102
261	Rutherfordium	Rf	104
262	Lawrencium	Lr	103
262	Dubnium	Db	105
264	Bohrium	Bh	107
266	Seaborgium	Sg	106
268	Meitnerium	Mt	109
272	Roentgenium	Rg	111
277	Hassium	Hs	108

3.2 Methods to Prepare Buffer Solutions

100 mM phosphoric acid (sodium) buffer solution (pH=2.1)
Sodium dihydrogen phosphate dihydrate (M. W. =156.01) 50 mmol (7.8 g)
Phosphoric acid (85 %, 14.7 mol/L) 50 mmol (3.4 mL)
Add water to make up to 1L.

10 mM phosphoric acid (sodium) buffer solution (pH=2.6)
Sodium dihydrogen phosphate dihydrate (M. W. =156.01) 5 mmol (0.78 g)
Phosphoric acid (85 %, 14.7 mol/L) 5 mmol (0.34 mL)
Add water to make up to 1 L.
(Alternatively, dilute 100 mM phosphoric acid (sodium) buffer solution (pH=2.1) ten times.)

50 mM phosphoric acid (sodium) buffer solution (pH=2.8)
Sodium dihydrogen phosphate dihydrate (M. W. =156.01) 40 mmol (6.24 g)
Phosphoric acid (85 %, 14.7 mol/L) 10 mmol (0.68 mL)
Add water to make up to 1 L.

100 mM phosphoric acid (sodium) buffer solution (pH=6.8)
Sodium dihydrogen phosphate dihydrate (M. W. =156.01) 50 mmol (7.8 g)
Sodium dihydrogen phosphate 12 - hydrate (M. W. =358.14) 50 mmol (17.9 g)
Add water to make up to 1 L.

10 mM phosphoric acid (sodium)buffer solution (pH=6.9)
Sodium dihydrogen phosphate dihydrate (M. W. =156.01) 5 mmol (0.78 g)
Sodium dihydrogen phosphate 12 - hydrate (M. W. =358.14) 5 mmol (1.79 g)
Add water to make up to 1 L.
(Alternatively, dilute 100 mM phosphoric acid (sodium) buffer solution (pH=6.8) ten times.)

20 mM citric acid (sodium) buffer solution (pH=3.1)
Citric Acid Monohydrate (M. W. =210.14) 16.7 mmol (3.51 g)
Trisodium Citrate Dihydrate (M. W. =294.10) 3.3 mmol (0.97 g)
Add water to make up to 1 L.

20 mM citric acid (sodium) buffer solution (pH=4.6)
Citric Acid Monohydrate (M. W. =210.14) 10 mmol (2.1 g)
Trisodium Citrate Dihydrate (M. W. =294.10) 10 mmol (2.94 g)
Add water to make up to 1 L.

10 mM tartaric acid (sodium) buffer solution (pH=2.9)
Tartaric acid (M. W. =150.09) 7.5 mmol (1.13 g)
Sodium tartrate dihydrate (M. W. =230.08) 2.5 mmol (0.58 g)
Add water to make up to 1 L.

10 mM tartaric acid (sodium) buffer solution (pH=4. 2)	
Tartaric acid (M. W. =150. 09) 2. 5 mmol (0. 375 g)	
Sodium tart rate dihydrate (M. W. =230. 08) 7. 5 mmol (1. 726 g)	
Add water to make up to 1 L.	

20mM (acetic acid) ethanolamine buffer solution (pH=9. 6)
Monoethanolamine (M. W. =61. 87, d=1. 017) 20 mmol (1. 22 mL)
Acetic acid (glacial acetic acid, 17. 4 mol/L) 10 mmol (0. 575 mL)
Add water to make up to 1 L.

100 mM aceticacid (sodium) buffer solution (pH=4. 7)
Acetic acid (glacial acetic acid) (99. 5 %, 17. 4 mol/L) 50 mmol (2. 87 mL)
Sodium acetate trihydrate (M. W. =136. 08) 50 mmol (6. 80 g)
Add water to make up to 1 L.

100 mM boric acid (potassium) buffer solution (pH=9. 1)
Boric acid (M. W. =61. 83) 100 mmol (6. 18 g)
Potassium hydroxide (M. W. =56. 11) 50 mmol (2. 81 g)
Add water to make up to 1 L.

100 mM boric acid (sodium) buffer solution (pH=9. 1)
Boric acid (M. W. =61. 83) 100 mmol (6. 18 g)
Sodium hydroxide (M. W. =40. 00) 50 mmol (2. 00 g)
Add water to make up to 1 L.

3.3 Standard Electrode Potential

Cathode (Reduction) Half-Reaction	Standard Potential E° (volts)
$Li^+ (aq) + e^- \longrightarrow Li(s)$	$-3. 04$
$Cu^+ (aq) + e^- \longrightarrow Cu(s)$	$-3. 026$
$K^+ (aq) + e^- \longrightarrow K(s)$	$-2. 92$
$Ca^{2+} (aq) + 2e^- \rightarrow Ca(s)$	$-2. 76$
$Na^+ (aq) + e^- \longrightarrow Na(s)$	$-2. 71$
$Mg^+ (aq) + 2e^- \longrightarrow Mg(s)$	$-2. 7$
$Mg^{2+} (aq) + 2e^- \longrightarrow Mg(s)$	$-2. 38$
$Al^{3+} (aq) + 3e^- \longrightarrow Al(s)$	$-1. 66$
$Mn^{2+} (aq) + 2e^- \longrightarrow Mn(s)$	$-1. 185$
$2H_2O(l) + 2e^- \longrightarrow H_2(g) + 2OH^- (aq)$	$-0. 83$
$Zn^{2+} (aq) + 2e^- \longrightarrow Zn(s)$	$-0. 762$
$Cr^{3+} (aq) + 3e^- \longrightarrow Cr(s)$	$-0. 74$
$Ag_2S(aq) + 2e^- \longrightarrow 2Ag(s) + S^{2-} (aq)$	$-0. 691$

(**Continued**)

Cathode (Reduction) Half-Reaction	Standard Potential $E°$ (volts)
$Fe^{2+}(aq)+2e^- \longrightarrow Fe(s)$	-0.44
$S(aq)+2e^- \longrightarrow S^{2-}(aq)$	-0.428
$PbSO_4(aq)+2e^- \longrightarrow Pb(s)+SO_4^{2-}(aq)$	-0.359
$Cd^{2+}(aq)+2e^- \longrightarrow Cd(s)$	-0.4
$Ni^{2+}(aq)+2e^- \longrightarrow Ni(s)$	-0.23
$AgI(aq)+e^- \longrightarrow Ag(s)+I^-(aq)$	-0.152
$Sn^{2+}(aq)+2e^- \longrightarrow Sn(s)$	-0.14
$Pb^{2+}(aq)+2e^- \longrightarrow Pb(s)$	-0.13
$SO_4^{2-}(aq)+H_2(aq)+2e^- \longrightarrow SO_3^{2-}(aq)+2OH^-(aq)$	-0.93
$Cr^{2+}(aq)+2e^- \longrightarrow Cr(s)$	-0.0913
$Cr^{3+}(aq)+e^- \longrightarrow Cr^{2+}(aq)$	-0.0407
$Fe^{3+}(aq)+3e^- \longrightarrow Fe(s)$	-0.037
$Ag_2S(aq)+2H^+(aq)+2e^- \longrightarrow 2Ag(s)+H_2S(aq)$	-0.037
$2H^+(aq)+2e^- \longrightarrow H_2(g)$	0
$AgBr(aq)+e^- \longrightarrow Ag(s)+Br^-(aq)$	0.0713
$S(aq)+2H^+(aq)+2e^- \longrightarrow H_2S(aq)$	0.142
$Sn^{4+}(aq)+2e^- \longrightarrow Sn^{2+}(aq)$	0.15
$Cu^{2+}(aq)+e^- \longrightarrow Cu^+(aq)$	0.16
$ClO_4^-(aq)+H_2O(l)+2e^- \longrightarrow ClO_3^-(aq)+2OH^-(aq)$	0.17
$AgCl(s)+e^- \longrightarrow Ag(s)+Cl^-(aq)$	0.22
$Cu^{2+}(aq)+2e^- \longrightarrow Cu(s)$	0.34
$ClO_3^-(aq)+H_2O(l)+2e^- \longrightarrow ClO_2^-(aq)+2OH^-(aq)$	0.35
$IO^-(aq)+H_2O(l)+2e^- \longrightarrow I^-(aq)+2OH^-(aq)$	0.49
$Cu^+(aq)+e^- \longrightarrow Cu(s)$	0.52
$I_2(s)+2e^- \longrightarrow 2I^-(aq)$	0.54
$AgNO_2(l)+e^- \longrightarrow Ag(s)+2\ NO_2^-(aq)$	0.564
$ClO_2^-(aq)+H_2O(l)+2e^- \longrightarrow ClO^-(aq)+2OH^-(aq)$	0.59
$O_2(g)+2H^+(aq)+2e^- \longrightarrow 2\ H_2O_2(l)$	0.695
$Fe^{3+}(aq)+e^- \longrightarrow Fe^{2+}(aq)$	0.77
$AgF(aq)+e^- \longrightarrow Ag(s)+F^-(aq)$	0.779
$Hg_2^{2+}(aq)+2e^- \longrightarrow 2Hg(l)$	0.8

(**Continued**)

Cathode (Reduction) Half-Reaction	Standard Potential $E°$ (volts)
$Ag^+(aq)+e^- \longrightarrow Ag(s)$	0.8
$Hg^{2+}(aq)+2e^- \longrightarrow Hg(l)$	0.85
$ClO^-(aq)+H_2O(l)+2e^- \longrightarrow Cl^-(aq)+2OH^-(aq)$	0.9
$2Hg^{2+}(aq)+2e^- \longrightarrow Hg_2^{2+}(aq)$	0.9
$NO_3^-(aq)+4H^+(aq)+3e^- \longrightarrow NO(g)+2H_2O(l)$	0.96
$Br_2(l)+2e^- \longrightarrow 2Br^-(aq)$	1.07
$Br_2(aq)+2e^- \longrightarrow 2Br^-(aq)$	1.0873
$Pt^{2+}(aq)+2e^- \longrightarrow Pt(s)$	1.18
$O_2(g)+4H^+(aq)+4e^- \longrightarrow 2H_2O(l)$	1.229
$Cr_2O_7^{2-}(aq)+14H^+(aq)+6e^- \longrightarrow 2Cr^{3+}(aq)+7H_2O(l)$	1.33
$Cl_2(g)+2e^- \longrightarrow 2Cl^-(aq)$	1.36
$Ce^{4+}(aq)+e^- \longrightarrow Ce^{3+}(aq)$	1.44
$MnO_4^-(aq)+8H^+(aq)+5e^- \longrightarrow Mn^{2+}(aq)+4H_2O(l)$	1.49
$Mn^{3+}(aq)+e^- \longrightarrow Mn^{2+}(aq)$	1.5415
$H_2O_2(aq)+2H^+(aq)+2e^- \longrightarrow 2H_2O(l)$	1.78
$Ag^{3+}(aq)+e^- \longrightarrow Ag^{2+}(aq)$	1.8
$Co^{3+}(aq)+e^- \longrightarrow Co^{2+}(aq)$	1.82
$Ag^{2+}(aq)+e^- \longrightarrow Ag^+(aq)$	1.98
$S_2O_8^{2-}(aq)+2e^- \longrightarrow 2SO_4^{2-}(aq)$	2.01
$O_3(g)+2H^+(aq)+2e^- \longrightarrow O_2(g)+H_2O(l)$	2.07
$Cu^{3+}(aq)+e^- \longrightarrow Cu^{2+}(aq)$	2.4
$F_2(g)+2e^- \longrightarrow 2F^-(aq)$	2.87
$F_2(g)+2H^++2e^- \longrightarrow 2HF(l)$	3.053

3.4　Solubility Product Constant

Compound	Formula	K_{sp} (25 ℃)
Aluminium hydroxide	$Al(OH)_3$	3×10^{-34}
Aluminium phosphate	$AlPO_4$	9.84×10^{-21}
Barium bromate	$Ba(BrO_3)_2$	2.43×10^{-4}
Barium carbonate	$BaCO_3$	2.58×10^{-9}

(**Continued**)

Compound	Formula	K_{sp} (25 ℃)
Barium chromate	$BaCrO_4$	1.17×10^{-10}
Barium fluoride	BaF_2	1.84×10^{-7}
Barium hydroxide octahydrate	$Ba(OH)_2 \times 8H_2O$	2.55×10^{-4}
Barium iodate	$Ba(IO_3)_2$	4.01×10^{-9}
Barium iodate monohydrate	$Ba(IO_3)_2 \times H_2O$	1.67×10^{-9}
Barium molybdate	$BaMoO_4$	3.54×10^{-8}
Barium nitrate	$Ba(NO_3)_2$	4.64×10^{-3}
Barium selenate	$BaSeO_4$	3.40×10^{-8}
Barium sulfate	$BaSO_4$	1.08×10^{-10}
Barium sulfite	$BaSO_3$	5.0×10^{-10}
Beryllium hydroxide	$Be(OH)_2$	6.92×10^{-22}
Bismuth arsenate	$BiAsO_4$	4.43×10^{-10}
Bismuth iodide	BiI	7.71×10^{-19}
Cadmium arsenate	$Cd_3(AsO_4)_2$	2.2×10^{-33}
Cadmium carbonate	$CdCO_3$	1.0×10^{-12}
Cadmium fluoride	CdF_2	6.44×10^{-3}
Cadmium hydroxide	$Cd(OH)_2$	7.2×10^{-15}
Cadmium iodate	$Cd(IO_3)_2$	2.5×10^{-8}
Cadmium oxalate trihydrate	$CdC_2O_4 \times 3H_2O$	1.42×10^{-8}
Cadmium phosphate	$Cd_3(PO_4)_2$	2.53×10^{-33}
Cadmium sulfide	CdS	1×10^{-27}
Caesium perchlorate	$CsClO_4$	3.95×10^{-3}
Caesium periodate	$CsIO_4$	5.16×10^{-6}
Calcium carbonate(*aragonite*)	$CaCO_3$	6.0×10^{-9}
Calcium carbonate(*calcite*)	$CaCO_3$	3.36×10^{-9}
Calcium fluoride	CaF_2	3.45×10^{-11}
Calcium hydroxide	$Ca(OH)_2$	5.02×10^{-6}
Calcium iodate	$Ca(IO_3)_2$	6.47×10^{-6}
Calcium iodate hexahydrate	$Ca(IO_3)_2 \times 6H_2O$	7.10×10^{-7}
Calcium molybdate	$CaMoO$	1.46×10^{-8}
Calcium oxalate monohydrate	$CaC_2O_4 \times H_2O$	2.32×10^{-9}

(**Continued**)

Compound	Formula	$K_{sp}(25\ ℃)$
Calcium phosphate	$Ca_3(PO_4)_2$	$2.07×10^{-33}$
Calcium sulfate	$CaSO_4$	$4.93×10^{-5}$
Calcium sulfate dihydrate	$CaSO_4×2H_2O$	$3.14×10^{-5}$
Calcium sulfate hemihydrate	$CaSO_4×0.5H_2O$	$3.1×10^{-7}$
Cobalt(II) arsenate	$Co_3(AsO_4)_2$	$6.80×10^{-29}$
Cobalt(II) carbonate	$CoCO_3$	$1.0×10^{-10}$
Cobalt(II) hydroxide(*blue*)	$Co(OH)_2$	$5.92×10^{-15}$
Cobalt(II) iodate dihydrate	$Co(IO_3)_2×2H_2O$	$1.21×10^{-2}$
Cobalt(II) phosphate	$Co_3(PO_4)_2$	$2.05×10^{-35}$
Cobalt(II) sulfide(*alpha*)	CoS	$5×10^{-22}$
Cobalt(II) sulfide(*beta*)	CoS	$3×10^{-26}$
Copper(I) bromide	$CuBr$	$6.27×10^{-9}$
Copper(I) chloride	$CuCl$	$1.72×10^{-7}$
Copper(I) cyanide	$CuCN$	$3.47×10^{-20}$
Copper(I) hydroxide	Cu_2O	$2×10^{-15}$
Copper(I) iodide	CuI	$1.27×10^{-12}$
Copper(I) thiocyanate	$CuSCN$	$1.77×10^{-13}$
Copper(II) arsenate	$Cu_3(AsO_4)_2$	$7.95×10^{-36}$
Copper(II) hydroxide	$Cu(OH)_2$	$4.8×10^{-20}$
Copper(II) iodate monohydrate	$Cu(IO_3)_2×H_2O$	$6.94×10^{-8}$
Copper(II) oxalate	CuC_2O_4	$4.43×10^{-10}$
Copper(II) phosphate	$Cu_3(PO_4)_2$	$1.40×10^{-37}$
Copper(II) sulfide	CuS	$8×10^{-37}$
Europium(III) hydroxide	$Eu(OH)_3$	$9.38×10^{-27}$
Gallium(III) hydroxide	$Ga(OH)_3$	$7.28×10^{-36}$
Iron(II) carbonate	$FeCO_3$	$3.13×10^{-11}$
Iron(II) fluoride	FeF_2	$2.36×10^{-6}$
Iron(II) hydroxide	$Fe(OH)_2$	$4.87×10^{-17}$
Iron(II) sulfide	FeS	$8×10^{-19}$
Iron(III) hydroxide	$Fe(OH)_3$	$2.79×10^{-39}$
Iron(III) phosphate dihydrate	$FePO_4×2H_2O$	$9.91×10^{-16}$

(**Continued**)

Compound	Formula	K_{sp} (25 ℃)
Lanthanum iodate	$La(IO_3)_3$	7.50×10^{-12}
Lead(II) bromide	$PbBr_2$	6.60×10^{-6}
Lead(II) carbonate	$PbCO_3$	7.40×10^{-14}
Lead(II) chloride	$PbCl_2$	1.70×10^{-5}
Lead(II) chromate	$PbCrO_4$	3×10^{-13}
Lead(II) fluoride	PbF_2	3.3×10^{-8}
Lead(II) hydroxide	$Pb(OH)_2$	1.43×10^{-20}
Lead(II) iodate	$Pb(IO_3)_2$	3.69×10^{-13}
Lead(II) iodide	PbI_2	9.8×10^{-9}
Lead(II) oxalate	PbC_2O_4	8.5×10^{-9}
Lead(II) selenate	$PbSeO_4$	1.37×10^{-7}
Lead(II) sulfate	$PbSO_4$	2.53×10^{-8}
Lead(II) sulfide	PbS	3×10^{-28}
Lithium carbonate	Li_2CO_3	8.15×10^{-4}
Lithium fluoride	LiF	1.84×10^{-3}
Lithium phosphate	Li_3PO_4	2.37×10^{-4}
Magnesium ammonium phosphate	$MgNH_4PO_4$	3×10^{-13}
Magnesium carbonate	$MgCO_3$	6.82×10^{-6}
Magnesium carbonate pentahydrate	$MgCO_3 \times 5H_2O$	3.79×10^{-6}
Magnesium carbonate trihydrate	$MgCO_3 \times 3H_2O$	2.38×10^{-6}
Magnesium fluoride	MgF_2	5.16×10^{-11}
Magnesium hydroxide	$Mg(OH)_2$	5.61×10^{-12}
Magnesium oxalate dihydrate	$MgC_2O_4 \times 2H_2O$	4.83×10^{-6}
Magnesium phosphate	$Mg_3(PO_4)_2$	1.04×10^{-24}
Manganese(II) carbonate	$MnCO_3$	2.24×10^{-11}
Manganese(II) hydroxide	$Mn(OH)_2$	2×10^{-13}
Manganese(II) iodate	$Mn(IO_3)_2$	4.37×10^{-7}
Manganese(II) oxalate dihydrate	$MnC_2O_4 \times 2H_2O$	1.70×10^{-7}
Manganese(II) sulfide(*green*)	MnS	3×10^{-14}
Manganese(II) sulfide(*pink*)	MnS	3×10^{-11}
Mercury(I) bromide	Hg_2Br_2	6.40×10^{-23}

Compound	Formula	$K_{sp}(25\ ℃)$
Mercury(I) carbonate	Hg_2CO_3	$3.6×10^{-17}$
Mercury(I) chloride	Hg_2Cl_2	$1.43×10^{-18}$
Mercury(I) fluoride	Hg_2F_2	$3.10×10^{-6}$
Mercury(I) iodide	Hg_2I_2	$5.2×10^{-29}$
Mercury(I) oxalate	$Hg_2C_2O_4$	$1.75×10^{-13}$
Mercury(I) sulfate	Hg_2SO_4	$6.5×10^{-7}$
Mercury(I) thiocyanate	$Hg_2(SCN)_2$	$3.2×10^{-20}$
Mercury(II) bromide	$HgBr_2$	$6.2×10^{-20}$
Mercury(II) hydroxide	HgO	$3.6×10^{-26}$
Mercury(II) iodide	HgI_2	$2.9×10^{-29}$
Mercury(II) sulfide(*black*)	HgS	$2×10^{-53}$
Mercury(II) sulfide(*red*)	HgS	$2×10^{-54}$
Neodymiumcarbonate	$Nd_2(CO_3)_3$	$1.08×10^{-33}$
Nickel(II) carbonate	$NiCO_3$	$1.42×10^{-7}$
Nickel(II) hydroxide	$Ni(OH)_2$	$5.48×10^{-16}$
Nickel(II) iodate	$Ni(IO_3)_2$	$4.71×10^{-5}$
Nickel(II) phosphate	$Ni_3(PO_4)_2$	$4.74×10^{-32}$
Nickel(II) sulfide(*alpha*)	NiS	$4×10^{-20}$
Nickel(II) sulfide(*beta*)	NiS	$1.3×10^{-25}$
Palladium(II) thiocyanate	$Pd(SCN)_2$	$4.39×10^{-23}$
Potassium hexachloroplatinate	K_2PtCl_6	$7.48×10^{-6}$
Potassium perchlorate	$KClO_4$	$1.05×10^{-2}$
Potassium periodate	KIO_4	$3.71×10^{-4}$
Praseodymium hydroxide	$Pr(OH)_3$	$3.39×10^{-24}$
Radium iodate	$Ra(IO_3)_2$	$1.16×10^{-9}$
Radium sulfate	$RaSO_4$	$3.66×10^{-11}$
Rubidium perchlorate	$RuClO_4$	$3.00×10^{-3}$
Scandium fluoride	ScF_3	$5.81×10^{-24}$
Scandium hydroxide	$Sc(OH)_3$	$2.22×10^{-31}$
Silver(I) acetate	$AgCH_3COO$	$1.94×10^{-3}$
Silver(I) arsenate	Ag_3AsO_4	$1.03×10^{-22}$

(**Continued**)

Compound	Formula	K_{sp} (25 ℃)
Silver(I) bromate	$AgBrO_3$	5.38×10^{-5}
Silver(I) bromide	$AgBr$	5.35×10^{-13}
Silver(I) carbonate	Ag_2CO_3	8.46×10^{-12}
Silver(I) chloride	$AgCl$	1.77×10^{-10}
Silver(I) chromate	Ag_2CrO_4	1.12×10^{-12}
Silver(I) cyanide	$AgCN$	5.97×10^{-17}
Silver(I) iodate	$AgIO_3$	3.17×10^{-8}
Silver(I) iodide	AgI	8.52×10^{-17}
Silver(I) oxalate	$Ag_2C_2O_4$	5.40×10^{-12}
Silver(I) phosphate	Ag_3PO_4	8.89×10^{-17}
Silver(I) sulfate	Ag_2SO_4	1.20×10^{-5}
Silver(I) sulfide	Ag_2S	8×10^{-51}
Silver(I) sulfite	Ag_2SO_3	1.50×10^{-14}
Silver(I) thiocyanate	$AgSCN$	1.03×10^{-12}
Strontium arsenate	$Sr_3(AsO_4)_2$	4.29×10^{-19}
Strontium carbonate	$SrCO_3$	5.60×10^{-10}
Strontium fluoride	SrF_2	4.33×10^{-9}
Strontium iodate	$Sr(IO_3)_2$	1.14×10^{-7}
Strontium iodate hexahydrate	$Sr(IO_3)_2 \times 6H_2O$	4.55×10^{-7}
Strontium iodate monohydrate	$Sr(IO_3)_2 \times H_2O$	3.77×10^{-7}
Strontium oxalate	SrC_2O_4	5×10^{-8}
Strontium sulfate	$SrSO_4$	3.44×10^{-7}
Thallium(I) bromate	$TlBrO_3$	1.10×10^{-4}
Thallium(I) bromide	$TlBr$	3.71×10^{-6}
Thallium(I) chloride	$TlCl$	1.86×10^{-4}
Thallium(I) chromate	Tl_2CrO_4	8.67×10^{-13}
Thallium(I) hydroxide	$Tl(OH)_3$	1.68×10^{-44}
Thallium(I) iodate	$TlIO_3$	3.12×10^{-6}
Thallium(I) iodide	TlI	5.54×10^{-8}
Thallium(I) sulfide	Tl_2S	6×10^{-22}
Thallium(I) thiocyanate	$TlSCN$	1.57×10^{-4}

(Continued)

Compound	Formula	K_{sp} (25 ℃)
Tin(II) hydroxide	$Sn(OH)_2$	5.45×10^{-27}
Yttrium carbonate	$Y_2(CO_3)_3$	1.03×10^{-31}
Yttrium fluoride	YF_3	8.62×10^{-21}
Yttriumhydroxide	$Y(OH)_3$	1.00×10^{-22}
Yttrium iodate	$Y(IO_3)_3$	1.12×10^{-10}
Zinc arsenate	$Zn_3(AsO_4)_2$	2.8×10^{-28}
Zinc carbonate	$ZnCO_3$	1.46×10^{-10}
Zinc carbonate monohydrate	$ZnCO_3 \times H_2O$	5.42×10^{-11}
Zinc fluoride	ZnF	3.04×10^{-2}
Zinc hydroxide	$Zn(OH)_2$	3×10^{-17}
Zinc iodatedihydrate	$Zn(IO_3)_2 \times 2H_2O$	4.1×10^{-6}
Zinc oxalate dihydrate	$ZnC_2O_4 \times 2H_2O$	1.38×10^{-9}
Zinc selenide	$ZnSe$	3.6×10^{-26}
Zinc selenite monohydrate	$ZnSeO_3 \times H_2O$	1.59×10^{-7}
Zinc sulfide(*alpha*)	ZnS	2×10^{-25}
Zinc sulfide(*beta*)	ZnS	3×10^{-23}

3.5 Organic Solvent Polarity

Compound	Formula	K_{sp} (25 ℃)
Aluminium hydroxide	$Al(OH)_3$	3×10^{-34}
Aluminium phosphate	$AlPO_4$	9.84×10^{-21}
Barium bromate	$Ba(BrO_3)_2$	2.43×10^{-4}
Barium carbonate	$BaCO_3$	2.58×10^{-9}
Barium chromate	$BaCrO_4$	1.17×10^{-10}
Barium fluoride	BaF_2	1.84×10^{-7}
Barium hydroxide octahydrate	$Ba(OH)_2 \times 8H_2O$	2.55×10^{-4}
Barium iodate	$Ba(IO_3)_2$	4.01×10^{-9}
Barium iodate monohydrate	$Ba(IO_3)_2 \times H_2O$	1.67×10^{-9}
Barium molybdate	$BaMoO_4$	3.54×10^{-8}
Barium nitrate	$Ba(NO_3)_2$	4.64×10^{-3}
Barium selenate	$BaSeO_4$	3.40×10^{-8}

(**Continued**)

Compound	Formula	K_{sp} (25 ℃)
Barium sulfate	$BaSO_4$	1.08×10^{-10}
Barium sulfite	$BaSO_3$	5.0×10^{-10}
Beryllium hydroxide	$Be(OH)_2$	6.92×10^{-22}
Bismuth arsenate	$BiAsO_4$	4.43×10^{-10}
Bismuth iodide	BiI	7.71×10^{-19}
Cadmium arsenate	$Cd_3(AsO_4)_2$	2.2×10^{-33}
Cadmium carbonate	$CdCO_3$	1.0×10^{-12}
Cadmium fluoride	CdF_2	6.44×10^{-3}
Cadmium hydroxide	$Cd(OH)_2$	7.2×10^{-15}
Cadmium iodate	$Cd(IO_3)_2$	2.5×10^{-8}
Cadmium oxalate trihydrate	$CdC_2O_4 \times 3H_2O$	1.42×10^{-8}
Cadmium phosphate	$Cd_3(PO_4)_2$	2.53×10^{-33}
Cadmium sulfide	CdS	1×10^{-27}
Caesium perchlorate	$CsClO_4$	3.95×10^{-3}
Caesium periodate	$CsIO_4$	5.16×10^{-6}
Calcium carbonate(*aragonite*)	$CaCO_3$	6.0×10^{-9}
Calcium carbonate(*calcite*)	$CaCO_3$	3.36×10^{-9}
Calcium fluoride	CaF_2	3.45×10^{-11}
Calcium hydroxide	$Ca(OH)_2$	5.02×10^{-6}
Calcium iodate	$Ca(IO_3)_2$	6.47×10^{-6}
Calcium iodate hexahydrate	$Ca(IO_3)_2 \times 6H_2O$	7.10×10^{-7}
Calcium molybdate	$CaMoO$	1.46×10^{-8}
Calcium oxalate monohydrate	$CaC_2O_4 \times H_2O$	2.32×10^{-9}
Calcium phosphate	$Ca_3(PO_4)_2$	2.07×10^{-33}
Calcium sulfate	$CaSO_4$	4.93×10^{-5}
Calcium sulfate dihydrate	$CaSO_4 \times 2H_2O$	3.14×10^{-5}
Calcium sulfate hemihydrate	$CaSO_4 \times 0.5H_2O$	3.1×10^{-7}
Cobalt(II) arsenate	$Co_3(AsO_4)_2$	6.80×10^{-29}
Cobalt(II) carbonate	$CoCO_3$	1.0×10^{-10}
Cobalt(II) hydroxide(*blue*)	$Co(OH)_2$	5.92×10^{-15}
Cobalt(II) iodate dihydrate	$Co(IO_3)_2 \times 2H_2O$	1.21×10^{-2}

Compound	Formula	K_{sp} (25 ℃)
Cobalt(II) phosphate	$Co_3(PO_4)_2$	2.05×10^{-35}
Cobalt(II) sulfide(*alpha*)	CoS	5×10^{-22}
Cobalt(II) sulfide(*beta*)	CoS	3×10^{-26}
Copper(I) bromide	$CuBr$	6.27×10^{-9}
Copper(I) chloride	$CuCl$	1.72×10^{-7}
Copper(I) cyanide	$CuCN$	3.47×10^{-20}
Copper(I) hydroxide	Cu_2O	2×10^{-15}
Copper(I) iodide	CuI	1.27×10^{-12}
Copper(I) thiocyanate	$CuSCN$	$1.77 \times 10-13$
Copper(II) arsenate	$Cu_3(AsO_4)_2$	7.95×10^{-36}
Copper(II) hydroxide	$Cu(OH)_2$	4.8×10^{-20}
Copper(II) iodate monohydrate	$Cu(IO_3)_2 \times H_2O$	6.94×10^{-8}
Copper(II) oxalate	CuC_2O_4	4.43×10^{-10}
Copper(II) phosphate	$Cu_3(PO_4)_2$	1.40×10^{-37}
Copper(II) sulfide	CuS	8×10^{-37}
Europium(III) hydroxide	$Eu(OH)_3$	9.38×10^{-27}
Gallium(III) hydroxide	$Ga(OH)_3$	7.28×10^{-36}
Iron(II) carbonate	$FeCO_3$	3.13×10^{-11}
Iron(II) fluoride	FeF_2	2.36×10^{-6}
Iron(II) hydroxide	$Fe(OH)_2$	4.87×10^{-17}
Iron(II) sulfide	FeS	8×10^{-19}
Iron(III) hydroxide	$Fe(OH)_3$	2.79×10^{-39}
Iron(III) phosphate dihydrate	$FePO_4 \times 2H_2O$	9.91×10^{-16}
Lanthanum iodate	$La(IO_3)_3$	7.50×10^{-12}
Lead(II) bromide	$PbBr_2$	6.60×10^{-6}
Lead(II) carbonate	$PbCO_3$	7.40×10^{-14}
Lead(II) chloride	$PbCl_2$	1.70×10^{-5}
Lead(II) chromate	$PbCrO_4$	3×10^{-13}
Lead(II) fluoride	PbF_2	3.3×10^{-8}
Lead(II) hydroxide	$Pb(OH)_2$	1.43×10^{-20}
Lead(II) iodate	$Pb(IO_3)_2$	3.69×10^{-13}

Compound	Formula	K_{sp} (25 ℃)
Lead(II) iodide	PbI_2	9.8×10^{-9}
Lead(II) oxalate	PbC_2O_4	8.5×10^{-9}
Lead(II) selenate	$PbSeO_4$	1.37×10^{-7}
Lead(II) sulfate	$PbSO_4$	2.53×10^{-8}
Lead(II) sulfide	PbS	3×10^{-28}
Lithium carbonate	Li_2CO_3	8.15×10^{-4}
Lithium fluoride	LiF	1.84×10^{-3}
Lithium phosphate	Li_3PO_4	2.37×10^{-4}
Magnesium ammonium phosphate	$MgNH_4PO_4$	3×10^{-13}
Magnesium carbonate	$MgCO_3$	6.82×10^{-6}
Magnesium carbonate pentahydrate	$MgCO_3 \times 5H_2O$	3.79×10^{-6}
Magnesium carbonate trihydrate	$MgCO_3 \times 3H_2O$	2.38×10^{-6}
Magnesium fluoride	MgF_2	5.16×10^{-11}
Magnesium hydroxide	$Mg(OH)_2$	5.61×10^{-12}
Magnesium oxalate dihydrate	$MgC_2O_4 \times 2H_2O$	4.83×10^{-6}
Magnesium phosphate	$Mg_3(PO_4)_2$	1.04×10^{-24}
Manganese(II) carbonate	$MnCO_3$	2.24×10^{-11}
Manganese(II) hydroxide	$Mn(OH)_2$	2×10^{-13}
Manganese(II) iodate	$Mn(IO_3)_2$	4.37×10^{-7}
Manganese(II) oxalate dihydrate	$MnC_2O_4 \times 2H_2O$	1.70×10^{-7}
Manganese(II) sulfide(*green*)	MnS	3×10^{-14}
Manganese(II) sulfide(*pink*)	MnS	3×10^{-11}
Mercury(I) bromide	Hg_2Br_2	6.40×10^{-23}
Mercury(I) carbonate	Hg_2CO_3	3.6×10^{-17}
Mercury(I) chloride	Hg_2Cl_2	1.43×10^{-18}
Mercury(I) fluoride	Hg_2F_2	3.10×10^{-6}
Mercury(I) iodide	Hg_2I_2	5.2×10^{-29}
Mercury(I) oxalate	$Hg_2C_2O_4$	1.75×10^{-13}
Mercury(I) sulfate	Hg_2SO_4	6.5×10^{-7}
Mercury(I) thiocyanate	$Hg_2(SCN)_2$	3.2×10^{-20}
Mercury(II) bromide	$HgBr_2$	6.2×10^{-20}

(**Continued**)

Compound	Formula	$K_{sp}(25\ ℃)$
Mercury(II) hydroxide	HgO	$3.6×10^{-26}$
Mercury(II) iodide	HgI$_2$	$2.9×10^{-29}$
Mercury(II) sulfide(*black*)	HgS	$2×10^{-53}$
Mercury(II) sulfide(*red*)	HgS	$2×10^{-54}$
Neodymiumcarbonate	Nd$_2$(CO$_3$)$_3$	$1.08×10^{-33}$
Nickel(II) carbonate	NiCO$_3$	$1.42×10^{-7}$
Nickel(II) hydroxide	Ni(OH)$_2$	$5.48×10^{-16}$
Nickel(II) iodate	Ni(IO$_3$)$_2$	$4.71×10^{-5}$
Nickel(II) phosphate	Ni$_3$(PO$_4$)$_2$	$4.74×10^{-32}$
Nickel(II) sulfide(*alpha*)	NiS	$4×10^{-20}$
Nickel(II) sulfide(*beta*)	NiS	$1.3×10^{-25}$
Palladium(II) thiocyanate	Pd(SCN)$_2$	$4.39×10^{-23}$
Potassium hexachloroplatinate	K$_2$PtCl$_6$	$7.48×10^{-6}$
Potassium perchlorate	KClO$_4$	$1.05×10^{-2}$
Potassium periodate	KIO$_4$	$3.71×10^{-4}$
Praseodymium hydroxide	Pr(OH)$_3$	$3.39×10^{-24}$
Radium iodate	Ra(IO$_3$)$_2$	$1.16×10^{-9}$
Radium sulfate	RaSO$_4$	$3.66×10^{-11}$
Rubidium perchlorate	RuClO$_4$	$3.00×10^{-3}$
Scandium fluoride	ScF$_3$	$5.81×10^{-24}$
Scandium hydroxide	Sc(OH)$_3$	$2.22×10^{-31}$
Silver(I) acetate	AgCH$_3$COO	$1.94×10^{-3}$
Silver(I) arsenate	Ag$_3$AsO$_4$	$1.03×10^{-22}$
Silver(I) bromate	AgBrO$_3$	$5.38×10^{-5}$
Silver(I) bromide	AgBr	$5.35×10^{-13}$
Silver(I) carbonate	Ag$_2$CO$_3$	$8.46×10^{-12}$
Silver(I) chloride	AgCl	$1.77×10^{-10}$
Silver(I) chromate	Ag$_2$CrO$_4$	$1.12×10^{-12}$
Silver(I) cyanide	AgCN	$5.97×10^{-17}$
Silver(I) iodate	AgIO$_3$	$3.17×10^{-8}$
Silver(I) iodide	AgI	$8.52×10^{-17}$

(**Continued**)

Compound	Formula	K_{sp}(25 ℃)
Silver(I) oxalate	$Ag_2C_2O_4$	5.40×10^{-12}
Silver(I) phosphate	Ag_3PO_4	8.89×10^{-17}
Silver(I) sulfate	Ag_2SO_4	1.20×10^{-5}
Silver(I) sulfide	Ag_2S	8×10^{-51}
Silver(I) sulfite	Ag_2SO_3	1.50×10^{-14}
Silver(I) thiocyanate	$AgSCN$	1.03×10^{-12}
Strontium arsenate	$Sr_3(AsO_4)_2$	4.29×10^{-19}
Strontium carbonate	$SrCO_3$	5.60×10^{-10}
Strontium fluoride	SrF_2	4.33×10^{-9}
Strontium iodate	$Sr(IO_3)_2$	1.14×10^{-7}
Strontium iodate hexahydrate	$Sr(IO_3)_2\times6H_2O$	4.55×10^{-7}
Strontium iodate monohydrate	$Sr(IO_3)_2\times H_2O$	3.77×10^{-7}
Strontium oxalate	SrC_2O_4	5×10^{-8}
Strontium sulfate	$SrSO_4$	3.44×10^{-7}
Thallium(I) bromate	$TlBrO_3$	1.10×10^{-4}
Thallium(I) bromide	$TlBr$	3.71×10^{-6}
Thallium(I) chloride	$TlCl$	1.86×10^{-4}
Thallium(I) chromate	Tl_2CrO_4	8.67×10^{-13}
Thallium(I) hydroxide	$Tl(OH)_3$	1.68×10^{-44}
Thallium(I) iodate	$TlIO_3$	3.12×10^{-6}
Thallium(I) iodide	TlI	5.54×10^{-8}
Thallium(I) sulfide	Tl_2S	6×10^{-22}
Thallium(I) thiocyanate	$TlSCN$	1.57×10^{-4}
Tin(II) hydroxide	$Sn(OH)_2$	5.45×10^{-27}
Yttrium carbonate	$Y_2(CO_3)_3$	1.03×10^{-31}
Yttrium fluoride	YF_3	8.62×10^{-21}
Yttrium hydroxide	$Y(OH)_3$	1.00×10^{-22}
Yttrium iodate	$Y(IO_3)_3$	1.12×10^{-10}
Zinc arsenate	$Zn_3(AsO_4)_2$	2.8×10^{-28}
Zinc carbonate	$ZnCO_3$	1.46×10^{-10}
Zinc carbonate monohydrate	$ZnCO_3\times H_2O$	5.42×10^{-11}

(**Continued**)

Compound	Formula	$K_{sp}(25\ ℃)$
Zinc fluoride	ZnF	3.04×10^{-2}
Zinc hydroxide	$Zn(OH)_2$	3×10^{-17}
Zinc iodatedihydrate	$Zn(IO_3)_2\times2H_2O$	4.1×10^{-6}
Zinc oxalate dihydrate	$ZnC_2O_4\times2H_2O$	1.38×10^{-9}
Zinc selenide	ZnSe	3.6×10^{-26}
Zinc selenite monohydrate	$ZnSeO_3\times H_2O$	1.59×10^{-7}
Zinc sulfide(*alpha*)	ZnS	2×10^{-25}
Zinc sulfide(*beta*)	ZnS	3×10^{-23}

4 References

1. S. Mumtazuddin, S. K. Sinha. Inorganic Lab Manual. Atlantic Publishers and Distributors Pvt Ltd, 2017.

2. M. Nath. Inorganic Chemistry: A Laboratory Manual. Alpha Science, 2016.

3. Anonymous. Laboratory Manual of Inorganic Chemistry: One Hundred Topics in General, Qualitative and Quantitative Chemistry. Franklin Classics, 2018.

4. H. Biltz. Laboratory Methods of Inorganic Chemistry. Antique Reprints, 2016.

5. J. A. Beran. Laboratory Manual for Principles of General Chemistry. Wiley, 2014.

6. D. C. Harris, C. A. Lucy. Quantitative Chemical Analysis. Freeman, 2019.

7. D. C. Harris, Quantitative Chemical Analysis. Freeman, 2015.

8. T. Bergin, An Introduction to Data Analysis: Quantitative, Qualitative and Mixed Methods. SAGE Publications Ltd, 2018.

9. R. H. Hill Jr., D. C. Finster. Laboratory Safety for Chemistry Students. Wiley, 2016.

10. 刘冰,徐强. 无机及分析化学实验(第 2 版)[M]. 北京:化学工业出版社,2015.

11. 吴建中. 无机化学实验[M]. 北京:科学出版社,2019.

12. 王艳玮,马兆立. 分析化学实验[M]. 北京:化学工业出版社,2020.

13. 徐建强. 定量分析实验与技术[M]. 北京:高等教育出版社,2018.